Abstraction in Theory- Laws of Physical Transaction

Theory of Everything

Subhajit Ganguly

Subhajit Ganguly

V M A

V M A Publications

Science Books

First published in 2011

Copyright © Subhajit Ganguly 2010

ISBN : 13: 978-1475072495

DEDICATED TO

Abstraction – the language of nature

CONTENTS

1 Preface

Over some years now, a large part of the energies of the scientific community has been employed solely for finding a theory that will fit in all known happenings of the physical world. Various groups of scientists have tried to attack the problem from different ends. Some of these theories have been partly successful in explaining the known physical world. However none of these theories have been without shortcomings. Be it the much lauded String Theory or the Quantum Gravity postulation or any other such attempts towards arriving at a Theory of Everything, none have been proved to be foolproof. To say the least, nobody can deny that there is room for much improvement before we can even start thinking truly towards such a theory that would describe the known world satisfactorily and provide for a single basis of understanding the four forces in nature.

On top of that, we have the newly emerging problems of 'Dark Energy', 'Dark Matter' and the like. These realms are yet to be accepted by the scientific community officially, but nonetheless, they are most definitely at least a few parts of mysteries that remain unexplained. A good and effective Theory of Everything must aim towards explaining such mysteries too. Sadly, we have no theory as yet that fulfills these criteria.

From the dawn of civilization, human beings have tried to find out order in the chaotic world surrounding them. It has however never been easy to find a solution to explain a given system while being a part of that system. The best bet is to find out the most fundamental components within the system and building a theory round these. In other words, a theory that is able to describe the world in totality has to keep the number of basic postulates it depends upon to zero or near zero. Deductionism hits a dead end in this regard. On the other hand, abstraction as the starting point of building up a theory may be seen to be of fitting use. It would be much more than a new way of tackling the problem. Even abstract postulates do away with the shackles that bind our theories into the system and bar them from being total descriptions of the system. The abstraction we are talking about here may be defined as, "Postulation of non-postulation" or, in other words, "A system of postulation that gives equal weights to all possible solutions inside the system and favours none of such solutions over others."

Abstraction automatically gives rise to optimized solutions within the universal set of all possible solutions, as has been shown in this book. It is these optimized solutions that make up and drive the non-abstract parts of the world, while the non-optimized solutions remain 'hidden' from the material world, inside the abstract world.

Starting from a basis of no postulation, we build our theory. As we go on piling up possibilities, we come to a similar basis for understanding the four non-contact forces of nature known till date. The difference in ranges of these forces is explained from this basis in this book. Zero postulation or abstraction as the basis of theory synthesis allows us to explore even imaginary and chaotic non-favoured solutions as possibilities. With no postulation as the fundamental basis, we are thus able to pile up postulated results or favoured results, but not the other way round. We keep describing such implications of abstraction in this book. We deal with the abstraction of observable parameters involved in a given system (quantum, relativistic, chaotic, non-chaotic)and formulate a similar basis of understanding them. Scaling of observable

parameters in adequate ways is shown to unite the understanding of worlds of the great vastness of the universe and the minuteness of the sub-atomic realm. Finally, the mysteries involving 'dark energy' and 'dark matter' are uncovered using such an approach.

This book is a culmination of the good wishes of many individuals, without whom nothing would have been possible. I take this opportunity to thank them all from the bottom of my heart. Hope my endeavour does justice to all the good wishes and aspirations surrounding it.

SUBHAJIT GANGULY

21-3-2012

2 Physical Transactions

Key Ideas:

Considering transport or tendency of transport of physical entities from an initial to a final point, we come to a similar basis of understanding of various physical phenomena. The trajectory-behaviour of such transport represents the effect or field of influence. This way, we may explain cluster-formation in the universe, an expanding universe, etc. This may also lead to a similar basis for understanding the four non-contact forces of nature. Also, for different ranges of acceleration in the field formed in space-time, we have different properties of matter interacting. This may explain the difference in ranges of the various forces.

Introduction:

All phenomena we see around us are but a physical or imaginary transport of various quantities starting from the flow of gases, to formation of fields of gravity, magnetism, etc., photons moving n space-time, right to the formation of cyclones and hurricanes, occurrence of events may be considered to be simply the transaction of physical or abstract entities between two points—one 'initial' and the other 'final'. The transfer of these various quantities, however, may not always be considered simple, and the 'route' of the transfer may form a complex relation with what may be called an environment. These transactions, however complex, must be obeying some physical laws, though, and all of them must be measurable for all physical considerations. There must be a same set of physical and mathematical laws governing all such transactions in every frame of reference. Consequently, we may consider the same set of a single mathematical construct, a set of equations that can safely describe all these phenomena satisfactorily. That concerned set of equations may take different forms, however, in different sets of events, but there must be a singular basis for all these. In a nutshell, we must be able to conjure up a mathematical idea that can explain such transfers of all known physical entities.

Developments in the mathematics of 'Chaos Theory' have pointed towards statistical analysis not to be always 'deterministic'. Meaning which, we cannot always predict a single solution to a given transfer in consideration, if it tends into the chaotic region of prediction. This however does not restrict us in finding mathematical tools for describing all possible routes that can be taken up by a given system in reaching a final point from an initial one. Taking into account all such possible routes of transfer or transactions, the safety of predictions increases to a maximum.

Factors governing transport:

Let us consider, somewhat arbitrarily, the factor that will govern the transfer of a given physical entity from an initial point 'A' to a final point 'B'.

(F) be the quantity of flow that takes place between the points, which are at a distance (D) apart. Now, there can be a support (S) towards the flow and a resistance (R) opposing it. For example, considering the spread of a given quantity of isolated gas in space from one point to another, the kinetic energies of the individual molecules may be taken to constitute the support towards the flow and the force of attraction between the molecules constitute the concerned resistance. Finally, let the quantity (T) represent the time in which the flow or transfer, we are concerned with, between the points A and B is complete.

Now, all other physical factors remaining constant, we have, respectively,

$$F \propto \frac{1}{D}$$
$$\propto \frac{1}{R}$$
$$\propto T$$
$$\propto S$$

The flow we are concerned with must also depend upon another factor. This being the difference in concentrations (λ) of the related physical entity between the two points concerned. All other factors remaining constant,

$$\frac{F}{T} \propto \lambda$$

Thereby, combining all concerned relations, we get,

$$F \propto \frac{\lambda ST}{DR}$$

or,

$$F \propto \omega \frac{\lambda ST}{DR} \qquad \qquad ...(1);$$

where ω is a constant of proportionality.

Let us now consider the flow of an energy quantum, with frequency (ν) in a given direction in vacuum. The distance (D) of transport in that given direction can be considered to be (cT);

where (c) is the velocity of energy-quantum in vacuum and (T) is the time of transport.

For an empty environment in which the flow takes place, we may assume that the concerned difference in concentrations between the initial and the final points,

$$\lambda = h\nu$$

where h is Plank's constant.

Placing $D = cT$ and $\lambda = h\nu$ in equation (1), we get,

$$F \propto \frac{\omega h \nu S}{cR}$$

Again, as the energy-quantum moves from the initial to the final point completely, the concerned flow,

$$F = h\nu$$

Placing this value of F in the previous equation, we get,

$$\omega = c\left(\frac{R}{S}\right)$$

Assuming, the resistance against the concerned flow and the support towards it for the energy, i.e., considering $R = S$, we have,

$$\omega = c \qquad\qquad\qquad ...(2)$$

The value of the constant ω may therefore be replaced by the speed of light in vacuum, c, in the equations concerning transport of a given physical entity.

Such a transfer of any physical entity as described by equation (1), will continue until and unless the difference in concentrations concerned, i.e., λ becomes zero. Considering the example of heat transfer from a body at a higher temperature to one at a lower temperature, heat will continue to flow towards the colder body, until and unless the difference in concentrations, i.e., the difference in temperatures of the hotter body and the colder one becomes zero. Similarly, a given body will keep moving with uniform velocity in a straight line, until and unless there is an acceleration or retardation (support or resistance, respectively, as the case might be), in accordance with equation (1).

In general, the following may be postulated from equation (1):

1. Any given physical quantity flows from a region of higher concentration of it to a region of lower concentration of it, until the concentrations in both the regions are the same.

2. Support (S) towards the concerned flow being zero in a given direction, the rate of flow $\left(\frac{dF}{dT}\right)$ is zero, as there is practically no flow in that given direction.

3. Resistance (R) against the flow being zero, time taken for the completion of the concerned flow is zero. Thus, if no resistance is present against the occurrence of a given event, it occurs practically instantaneously.

4. Considering any given region as a collection of a finite or an infinite number of initial and final points, we can say that any given entity within the region tends to flow equally in all directions. However, support and resistance towards and against a flow in any given direction will decide the routes of such a transport.

Instability in a System:

The logical approach that has been undertaken in studying transfer of a given entity between points seems to have led us to what may be called super-stability. As the given physical entity tends to flow equally in all directions, it seems that the universe, as we know it, should have been highly stable, with almost equal quantities of all physical quantities spread equally everywhere. Both matter and energy should have been in equal or almost equal proportions everywhere. But this is not so. The very logic concerned may be made use of to explain this seeming anomaly.

.   Let us consider the example of a three-points isolated system. Let the points be 'A', 'B' and 'C'. Let A and B be material points, whereas, C be situated anywhere on the straight line joining A and B. The material parts of both A and B tends to move in all possible directions. These possible directions include the directions towards each other. Thus, at point C, for obvious reasons, an additional effect will be felt due to the tendency of material to flow from A to B and from B to A, as compared to all other directions.

The points A, B and C being considered parts of an isolated system and all three points being assumed fundamentally similar (with the only difference that A and B contain material, while C is empty), the factors R and S must be equal.

Placing this condition in equation (1) and replacing ω by c, for reasons stated earlier, we get,

$$\frac{F}{T} = c\frac{\lambda}{D}$$

D being considered the distance between A and C and x the distance between A and B (say), the distance between B and C is $D - x$.

The effect on C due to the material-point A can thus be written as,

$$\frac{F_A}{T_A} = c\frac{\lambda_A}{x}$$

Similarly, the effect on the empty point C due to the material-point B is,

$$\frac{F_B}{T_B} = c\frac{\lambda_B}{D - x};$$

where F_A and F_B are the respective values of flows towards the point C due to A and B, respectively. T_A and T_B are the respective values of time and λ_A and λ_B are the respective values of the differences in concentrations of the concerned entity between A and B.

Substituting x in the above two equations, we have,

$$c\frac{T_A\lambda_A}{F_A} = D - c\frac{T_B\lambda_B}{F_B} \qquad \text{...(3)}$$

Considering the points to be having equal factors, i.e., considering $\lambda_A = \lambda_B = \lambda$ (say), $F_A = F_B = F$ (say) and $T_A = T_B = T$ (say), equation (3) reduces to,

$$c\frac{T\lambda}{F} = D - c\frac{T\lambda}{F}$$

i.e.,

$$\frac{F}{T} = 2c\left(\frac{\lambda}{D}\right) \qquad\qquad ...(4)$$

Equation (4) describes fundamentally the effect (i.e., the flow F in time T) of two material-points having same factorial conditions regarding one or a number of entities. Considering a collection of such points and applying a statistical approach, the logistic equation (due to May, 1967) for $\left(\frac{F}{T}\right)$ can be written as,

$$2c\left(\frac{\lambda}{D}\right)_{t+1} = 2Kc\left(\frac{\lambda}{D}\right)_t\left[1 - 2c\left(\frac{\lambda}{D}\right)_t\right]$$

i.e.,

$$\left(\frac{\lambda}{D}\right)_{t+1} = K\left(\frac{\lambda}{D}\right)_t\left[1 - 2c\left(\frac{\lambda}{D}\right)_t\right] \qquad\qquad ...(5)$$

where K is a constant.

Also, the quadratic map (due to Lorentz, 1987) can be written as,

$$2c\left(\frac{\lambda}{D}\right)_{t+1} = K - \left(2c\frac{\lambda}{D}\right)_t^2$$

i.e.,

$$2c\left(\frac{\lambda}{D}\right)_{t+1} = K - 4c^2\left(\frac{\lambda}{D}\right)_t^2 \qquad\qquad ...(6)$$

All trajectories described by the quadratic map become asymptotic to $-\infty$ for
$K < -0.25$ and $K > 2$

As we deal with the flow of a given material entity towards one given point or the effects on a given point, the expression for the attractor for each such point can be written as,

$$\left(2c\frac{\lambda}{D}\right)^* = \left(1 - \frac{1}{K}\right) \qquad\qquad ...(7);$$

where $0 < K < B$.

$\left(2c\frac{\lambda}{D}\right)'$ is a point in the desired dimensional plot into which the trajectories seem to crowd. As we do not need to deal with more than one attractor or periodic point, the trajectories will tend to revisit only the attractor point concerned, to the desired level of accuracy of observations and calculations.

In equation (7), for $K \geq 3$, the trajectory behaviour becomes increasingly sensitive to the value of K. There are a few more points to be noted regarding the dependence of the trajectory behaviour on the values of K:

1. For $K \leq 1$, the attractor is a fixed point and has a value 0.
2. For $1 < K < 3$, the attractor is a fixed point and its value is > 0 but < 0.667.
3. For $3 \leq K \leq 3.57$, period doubling occurs, with the attractor consisting of $2, 4, 8,$ etc., periodic points as K increases within that range.
4. For $3.57 < K \leq 4$, we have the region of chaos, where the attractor can be erratic (chaotic with infinitely many points) or stable.

For all calculations, the desired conditions may be placed at the attractor. A trajectory never gets completely and exactly all the way into an attractor though, but only approaches it asymptotically. In the region of chaos, we apply the method of searching for windows or zones of K-values for which iterations from any initial conditions will produce the periodic attractor, instead of a chaotic one. For the logistic equation (5), the most common such zone lies at $K \approx 3.83$ and for the quadratic map (6), at $K \approx 1.76$.

Making use of equations $(4), (5), (6)$ and (7), the effect of material-points on their surroundings can be studied. Also, it is important to note that the presence of material-points ensures trajectories of all possible routes of the transfer of entities and these may be even chaotic in nature. The geometry around such

11

points must therefore depend upon the probabilities concerned of the given trajectories revisiting the attractor-point. Transport of any physical entity (or even if it tends to be transported) will therefore affect the geometry of the surroundings in a similar way. Thus, we are bound to have fundamentally similar trajectories in all cases of transfer of entities. This being true, an actual transport or a 'tendency' of transport of any physical entity and its effect may fundamentally be predicted in a single similar fashion. By selecting the initial and the final points or regions concerned effectively, in any given scenario, we must be able to describe the behaviour of the transport to the desired level of accuracy, by studying the trajectories of it. Choosing appropriate factors too should be necessarily sufficient in taking care of the desired level of accuracy in calculations.

The tendency of the material-points to flow towards each other, as discussed, must be sufficient to explain the absence of super-stability in the universe. Also, the non-material or empty points in the vicinity of the material-points, as discussed earlier, tend to behave just as material-points, thereby causing a change in the whole vicinity or environment concerned. This predicts the presence of 'fields' in the environment of material and non-material points, further nullifying the super-stability of the universe, explaining why we see the world the way we see it.

Using Lyapunov exponents for a transport as described by equation (4), and replacing $2c\left(\frac{\lambda}{D}\right)$ by a quantity $'\tau'$, we have,

$$\frac{d}{d\tau} f^n(\tau) = \frac{\delta n}{\delta o}$$

i.e.,

$$\frac{\delta n}{\delta o} = \prod_{i=1}^{n} f'(\tau_i) \qquad \qquad \dots (8)$$

$$b = \frac{1}{n} \log_e \left(\frac{\delta n}{\delta o}\right)$$

i.e.,

$$b = \frac{1}{n} \sum_{i=1}^{n-1} \log_e |f'(\tau_i)| \qquad \ldots (9);$$

where b is a constant (the local slope of all possible routes), and

$$\Psi = \lim_{n \to \infty} \frac{1}{n} \sum_{i=0}^{n-1} \log_e |f'(\tau_i)| \qquad \ldots (10);$$

where Ψ is a constant.

Using equations $(8), (9)$ and (10), the probabilistic plots of all possible routes of transport are to be found, for a given scenario. These plots, in turn, yield the description of the effect or field of the concerned transport or transaction. Depending upon the transport being considered, we may have the respective sort of field, viz., electromagnetic, gravitational, etc. These fields, however, being fundamentally similar probability plots of all possible routes of transactions of physical or imaginary entities must have a similar basis for comprehension. Thus, there must be a grand unification of all physical theories of transport.

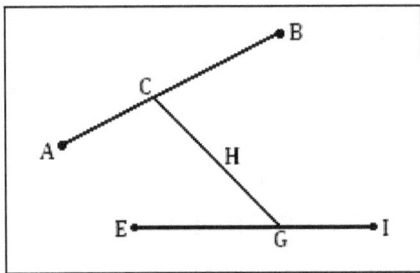

Fig. 1: Four material-points A, B, E and I, and two non-material points C and G in their respective vicinities

Let us consider, as shown in the figure above, four material-points A, B, E and I, and two non-material or 'empty' points C and G, anywhere on the straight-lines AB and EI respectively. Let the points A, B, E and I have equal factorial conditions of material transport (equal to F) difference in concentrations of material (equal to λ)

and concerned time (equal to T). Also, let AB and EI be having equal straight-line distances (equal to D). The effects due to A and B on the point C and those due to E and I on the point G are both, by equation (1),

$$\frac{F}{T} = 2\omega \left(\frac{\lambda}{D}\right)$$

ω being constant for the given system of transactions concerned.

Considering another point H anywhere on the straight-line distance between the points C and G, an effect must be felt at H too, due to the point C and G. By equation (1),this effect must be $\left(2\frac{F}{T}\right)$ or $\left(2\omega\frac{\lambda}{D} + 2\omega\frac{\lambda}{D}, \text{i.e.}, 4\omega\frac{\lambda}{D}\right)$. The points C and G, and as a cosequence the point H too, inspite of being considered empty points, are seen to behave just as material-points in the vicinity of material transactions. Effects of such a transaction affect non-material points and these non-material points in return can thus affect other material or non-material points. Thus, the points themselves that we consider, irrespective of being empty or containing material, behave as if being in a same kind of transport themselves as any material, in accordance with equation (1).

Therefore, we are to deal with not only real material transport, but transport of abstract entities too. Not only must a physical entity tend to move towards all points in its vicinity, the points that we consider, themselves must tend to be moving in all possible directions as we can see. The points themselves will have to be studied therefore by their trajectory behaviour as described by equations (5), (6) and (7) and their effects or fields of influence by equations (8), (9) and (10).

The amount of information required to describe the trajectory of a concerned plot to within an accuracy or length of measuring tool (ε) be (I_ε), say. We have,

$$I_\varepsilon = \sum_{i=1}^{N} P_i \log_2 \left(\frac{1}{P_i}\right);$$

where P_i represents the concerned relative frequencies or probabilities of individual observations. Writing logarithmically (arbitrarily), we have,

$$I_s = a + D_i \log_2 \left(\frac{1}{\varepsilon}\right);$$

where $'a'$ is constant, and

$$D_i = \lim_{\varepsilon \to 0} \left[\frac{I_s}{\log_2(u/\varepsilon)} - \frac{a}{\log_2(u/\varepsilon)}\right]$$

i.e.,

$$D_i = \lim_{\varepsilon \to 0} \frac{I_s}{\log_2(u/\varepsilon)} ;$$

$\frac{a}{\log_2(u/\varepsilon)}$ being sufficently small to be neglected (u represents the unit length of the original).The necessity of I_ε to fall within the desired value is absolute, barring which safety of predictions using the plot concerned is hampered. Ruler-length decreasing to 0, we have,

$$D_\varepsilon = \lim_{\varepsilon \to 0} \frac{\log N}{\log_2(1/\varepsilon)};$$

where D_ε is the concerned dimension of measurements.

Starting measurements relating to two regions, one initial and the other final, each such region may be further considered to be a collection of some other regions. The scaling operation is performed such that:

1. There is a finite number of sub-divisions.
2. Step 1 is repeated on each new facsimilie.

Thus, we have the dimension,

$$D = \frac{\log N}{\log_2 \left(\frac{1}{r}\right)};$$

where N is the number of facsimilies and r represents the scaling-ratio (i.e., 1/number of sub-divisions).

Also, for a unit length (u), $u = r^D N$

For irregular forms, however, the estimated unit length,
$L = \varepsilon N$

Inserting a constant of proportionality (a) in $N = \left(\frac{1}{r^D}\right)$, for going from unit length to a measured length, we get,

$$N = a\left(\frac{1}{r^D}\right)$$

Further, inserting new scaling length (ε) in place of (r), we have,

$$N = a\left(\frac{1}{\varepsilon^D}\right)$$

Thus, from the equations $(L = \varepsilon N)$ and $(N = a/\varepsilon^D)$, we have,
$L\varepsilon = a\varepsilon^{1-D}$

The value of D for an irregular line (eg., a Von Koch snowflake or a coastline) is between 1 and 2. The slope of a line fitted to all the points taken from the coastline may be estimated to be,
$m = 1 - D$

Using desired scaling-lengths, the necessary factorial conditions for a given transport may be estimated to the necessary level of accuracy, for performing calculations. Each of the quantities in equation (1) may have a given number of dimensions for a given level of acccuracy in predicting the routes of transaction. For the desired level of accuracy, any of the quantities (L_i) may be considered to depend upon (n) number of parameters (P_i). A desired level of accuracy for a given set of measurements can be achieved by knowing exactly the necessary parameters for the observations. In this respect, auto-correlation coefficient or lag-m serial correlation coefficient may be employed. This coefficient being,

$$A = \frac{\sum_{t=i}^{N-m}(P_t - \bar{P})(P_{t+m} - \bar{P})}{\sum_{t-1}^{N}(P_t - \bar{P})^2}$$

where $\bar{P}$ represents the mean value of the concerned paramater.

At a lag of 0, the co-ordinates of each plotted point are equal, auto-correlation coefficient being maximum. Auto-correlation decreases with increase in lag. By adequate accuracy in observations for a desired set of measurements, a necessary level of accuracy may thus be achieved.

Spillage in all possible directions may be prevented, for a given scenario, by the factorial conditions R or S for each of these directions.

An expanding Universe:

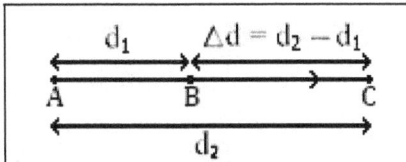

<u>Fig. 2</u>: Relative motion of a point from B to C, in vicinity of point A

Let us consider a point of reference A. Let another point move from point B to point C, with respect to point A. Let B and C be at distances d_1 and d_2 respectively from point A. Considering $S = R$ for the flow, and λ to be a constant (considering the movement of the concerned point from one 'empty' place to another), from equation (1), we have,

$$\frac{F}{T} \propto \frac{1}{D}$$

Thus, as the flow takes place from B towards C, as the distance between B and C ($\Delta d = d_2 - d_1$) decreases, the velocity of the flow $\left(\frac{F}{T}\right)$ increases. Thus, an observer at A will see that the velocity of the moving point goes on increasing as it keeps moving away from A. This is in consistency with Hubble's law ($v = Hr$), for an expanding universe.

As the point B keeps moving away from A towards point C, the effect of point A on point B goes on decreasing. The effect of the field around the reference point seems to lessen thus with distance, getting the points further from it to spread away with increasing velocity. The distance between the reference point and the moving point being sufficiently greater than the 'size' of the points concerned, this spreading away from the reference point assumes considerable importance for obvious reasons.

The Four Non-Contact Forces:

The four non-contact forces, viz., the electromagnetic, the weak, the strong and the gravitational forces, being considered the effects of transport of respective entities, may be treated as per equation (1). Equations $(8), (9)$ and (10) would yield the probabilistic plots or fields of all routes of actual transport or tendency of transport in each case. When the distance between the concerned initial and final points of transport become sufficiently large, however, than the 'size' of the points, the trajectories tend to become chaotic, thereby increasing the uncertainty of predictions. This uncertainty, in turn, depends upon the Lyapunov exponents (v). Moreover, there is a stretching or shrinking of a given direction according to the factor e^{vt}, according as v being positive or negative in that direction.

Let us suppose a system is characterized by a positive v, i.e., v_+ and its initial state is defined within a size ε. Then, in time T, the uncertainty in the co-ordinates concerned will have expanded to the size L of the attractor,

$$L \sim \varepsilon e^{v_+ T}$$

or,

$$L \sim \varepsilon e^{KT}$$

Either of these relations may be solved for the prediction-time,

$$T \sim \left(\frac{1}{v_+}\right) \log_e \left(\frac{L}{\varepsilon}\right)$$

or,

$$T \sim \left(\frac{1}{K}\right) \log_e \left(\frac{L}{\varepsilon}\right)$$

The prediction time, therefore, increases only logarithmically with the precision of the initial measurement. Thus, in such chaotic states, where the size of the concerned points becomes sufficiently small or large as compared to the concerned distance inbetween, only short-term predictions are possible.

Let a region in space-time be represented by,

$$s = (cT)^{4i};$$

where c is speed of light in vacuum and $i = \sqrt{-1}$.

Let us consider the introduction of a given physical property, viz., mass, charge, colour, etc., in this region. The vicinity s will be affected by this introduction and let this vicinity presently become,

$$s' = (cT')^{4i}$$

(If the size of the concerned source of disturbance is very small or very large as compared to its vicinity, chaotic states prevail, and no long term prediction will be possible.)

The change in space-time configuration,

$$\Delta s = (cT)^{4i} - (cT')^{4i}$$
$$= c^{4i}(T - T')^{4i}$$
$$= c^{4i}(\Delta T)^{4i} \qquad \text{(replacing } T - T' \text{ by } \Delta T)$$

This change in the space-time configuration will be caused as all the points in the vicinity of the source of the concerned physical property is affected by its field, as stated earlier. The velocity of this change in space-time configuration,

$$V = \frac{c^{4i}(\Delta T)^{4i}}{T}$$

and the acceleration related to this velocity will be,

$$a = \frac{c^{4i}(\Delta T)^{4i}}{T^2}$$

This acceleration will be felt by the concerned physical property in the vicinty of the source. As such, there will be a force for each of these properties concerned, viz., gravitational force for mass, electromagnetic for charge, colour-force for colour, etc. The value of time in the space-time is different for each of such interactions.

Therefore, the strength of the interacting force for each property is different too. In a system, where the resistance R and support S for a change are equal, i.e., for a neutral system, we have,

$$\frac{F}{T} = c\frac{\lambda}{D}$$

For the concerned field of influence, relating to the space-time vicinity,

$$F = (cT)^{4i}$$

Also,

$$D = (cT)^{4i};$$

as the distance of influence will be equal to the flow F.

Combining the last three relations, we get,

$$\frac{(cT)^{4i}}{T} = c\frac{\lambda}{(cT)^{4i}}$$

i.e.,

$$T = \frac{\lambda^{\frac{1}{8i-1}}}{c}$$

Also, for such a neutral system, the difference in concentrations λ, will be equal to the quantity of the property concerned, that is introduced, (say A); i.e.,

$$\lambda = A \text{ and } \Delta\lambda = \Delta A$$

Thus, we have,

$$T = \frac{A^{\frac{1}{8i-1}}}{c}$$

The change in configuration therefore is,

$$\Delta s = c^{4i}\,\Delta T^{4i}$$

$$= c^{4i}\left\{\frac{(\Delta A)^{\frac{1}{8i-1}}}{c}\right\}^{4i}$$

$$= (\Delta A)^{\frac{4i}{8i-1}}$$

Also, the acceleration in the vicinity of the source,

$$a = \frac{c^{4i}\left\{\frac{(\Delta A)^{\frac{1}{8i-1}}}{c}\right\}^{4i}}{\left(\frac{A^{\frac{1}{8i-1}}}{c}\right)^{2}}$$

i.e.,

$$a = c^{2}\frac{(\Delta A)^{\frac{4i}{8i-1}}}{(A)^{\frac{2}{8i-1}}} \qquad \qquad \ldots(11)$$

Thus, the force that will be felt due to a quantity A' of the property A in this field of acceleration is,

$$F = A'c^{2}\frac{(\Delta A)^{\frac{4i}{8i-1}}}{(A)^{\frac{2}{8i-1}}} \qquad \qquad \ldots(12)$$

The strength of this force, as stated earlier, will be difficult for each concerned physical property. However, there must be some similar basis of interaction with each of these strengths of forces. The trajectory behaviour for each of these fields of force may be studied using the Lyapunov exponents. For one-dimensional studies, the logistic equation similar to equation (5) may be used. For studying such trajectory behaviour in two-dimensional frame, a quadratic map similar to equation (6) may be used. Also, equation (7) gives the attractor for the trajectories concerned.

Also, the quantity of force forming due to the interaction of a property A' in a given acceleration-field $'a'$ may be considered to be different. Interaction of different properties of matter, viz., mass, charge, colour, etc., with the same acceleration-field may yield forces of different ranges. Thus, it may be considered that at different ranges of acceleration different properties of matter will interact, although, these may do so in an equivalent manner. Therefore, there must be different ranges of forces that have influence over different properties. This is consistent with available data. The types of forces that are seen arise due to the interaction of respective properties with an equivalent acceleration-field.

Conclusion:

Starting with studying actual transport or tendency to transport of entities, we have arrived at a conjecture, a theory that all physical phenomena may be explained through a single rule and thereby may be studied in a similar way. In each of these considerations, we take into account the initial and the final points of transport or flow. Choosing appropriate parameters for measurement and applying adequate observation, we can find the trajectories or the effects or fields of their behaviour. For a given set of calculations, we will have to take care of the desired level of the accuracy for the intended set of results. Even if we conveniently omit a few parameters or information regarding such transport, we may be able to reach our desired level of accuracy. Reaching towards the complete set of parameters though, will ensure moving towards complete accuracy of predictions, even if the related trajectories become chaotic. The prediction-time in chaotic regions becomes less though. Placing the desired direction of flow of the given physical entity along a given direction, we may find the extent of flow that spills into any of the other remaining directions.

We have also explained cluster-formation, or the coming together of physical entities in the universe, and the stability of such systems. An expanding universe may also be explained in a similar way.

Finally, considering transport or tendency of transport of different physical entities, we may arrive at the unification of the four non-contact forces of nature. A force arises due to the interaction of a given physical property with an acceleration-field, created due to the same property or an equivalent one. Different physical entities interact with different ranges of the acceleration-field and this may give rise to different ranges of the various forces.

3 A Few Implications of the Laws of Transactions, From the Abstraction Theory

Key Ideas:

Considering transport of light through spacetime, following the laws of physical transactions, it may be said that there must be a spreading effect on it. Over suitable distances from a source of light, an observer's perception is bound to be affected due to this spreading. In the following chapter, these effects on the reception of a signal, due to the spreading of light are studied. Experimental set-ups are devised to verify the actual angles of spread with their theoretical values. An experiment regarding the minimum distance between two disturbances for them to be distinguishable is also carried out. The energy quantum is also studied in a new light.

Introduction:

In accordance with the laws of physical transactions stated in the previous chapter, a light-signal emanating from a given source will spread. As such, reception of the signal at a suitable distance away from the source will be affected. An observer's perception of a given source is thus bound to be distorted from the original, if the source is at considerably large distances away (example: reception of light from other galaxies). Over smaller distances between the source and the observer, though, the effects of the spread of the light-signal may not be considerable however, being of negligibly small dimensions.

As an experimental set-up on an inter-galactic scale may not yet be possible, in order to examine the spread of light, we have to devise such a set-up which is possible and practical. This chapter also deals with such an experimental set-up, with which it is possible to verify the spread of different wavelengths of light. Theoretical values are tallied against the experimental ones. Further, the minimum distance of separation required for two waves to be distinguishable is studied through another experiment. Lastly, the energy-quantum is studied in accordance with the laws of physical transactions, with regards to the Theory of Abstraction.

Different Rates of Spread for Different Frequencies:

Notwithstanding the effect of spreading of photon-gas, the photons themselves must tend to expand, according to the last chapter. Light with different wavelengths must have a proportional rate of spreading.

Let Fig. 1 represent a simple projectile motion under an acceleration a.

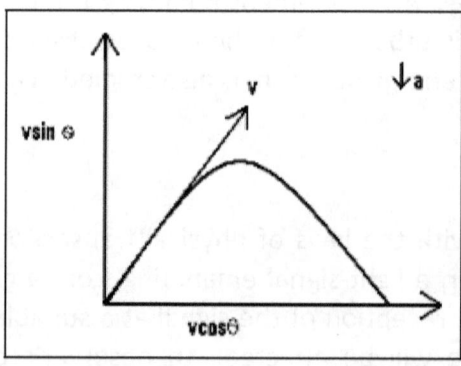

Fig. 1: Simple projectile (half the wave)

The range,
$$R = \frac{v^2 \sin 2\theta}{a} \quad \text{i. e., } a = \frac{v^2 \sin 2\theta}{R} \qquad \dots (1)$$

Time of flight,
$$t = \frac{2v \sin \theta}{a} \quad \text{i. e., } a = \frac{2v \sin \theta}{t} \qquad \dots (2)$$

The spreading of a single photon may be considered to be a wave in itself, which in turn may be considered to be a combination of two such projectile motions with a pseudo-acceleration (a), as shown in Fig. 2.

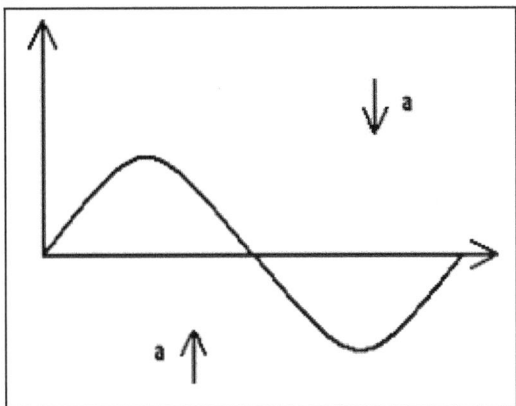

Fig. 2: Wave motion as a combination of two projectile motions

Again, the time of flight (t) of each projectile motion must be half the time-period (T) of the concerned radiation such that $t = \frac{1}{2}T$. Similarly, the range (R) of each projectile-motion may be considered to be half the wavelength (λ) of the radiation, such that, $R = \frac{\lambda}{2}$.

Dividing equation (1) by equation (2) and placing $t = \frac{1}{2}T$, $R = \frac{\lambda}{2}$ and $v = c$ (speed of light in vacuum), we get,

$$\cos\theta = \frac{\lambda}{Tc}$$

Considering lights with a difference in wavelengths ($\Delta\lambda$), over a distance x, for the difference in their angles of spreading ($\Delta\theta$), we may write,

$$\cos\Delta\theta = \frac{x}{\Delta Tc}$$

i.e.,

$$\cos\Delta\theta = \frac{\Delta\lambda}{x} \qquad\qquad \dots(3)$$

In order to test equation (3), we may send out lights of two given wavelengths (λ_1) and (λ_2) and then measure the angle ($\Delta\theta$) at which we get light of a different wavelength due to the mixing up of the two different wavelengths, for various distances (x).

The above experiment may be performed with two coherent sources of red and green lights in vacuum. The minimum distance at which yellow light is formed is to be measured. The experimental set-up is shown in fig. 3.

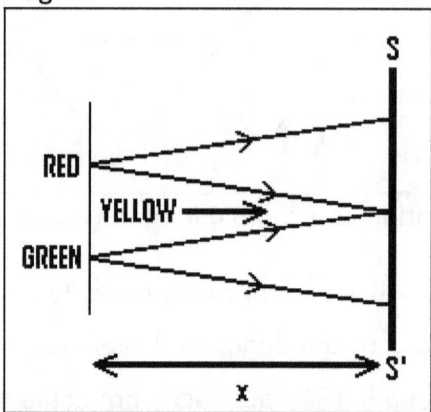

Fig. 3: Experimental set-up for measurement of x

As the red and green lights spread, they form yellow light at a distance x on a screen SS'. The sources being placed reasonably close to each other, x may be made to be comfortably small. Photographic-plates are used for precise measurements.

Using equation (3), we may write,

$$x = \frac{\Delta\lambda}{\cos\Delta\theta}$$

Colour	Wavelength (in nano-meters)	Difference in wavelengths $(\Delta\lambda)$(in meters)	Angle at which yellow-light is formed $(\Delta\theta)$	Distance (x)(in meters)
Red	680		$\cos^{-1}26.67 \times 10^{-9}$	6
Green	520	160×10^{-9}	$\cos^{-1}22.86 \times 10^{-9}$	7
Yellow	560			4
			$\cos^{-1}40\times10^{-9}$	

Table 1- Measurement of the distance and the angle of formation of yellow-light

For light frequency (ν) using similar treatment as earlier (reference fig. 1 and fig. 2), we can write,

$$a = 2c\nu\sin 2\theta \qquad \text{...}(4)$$

and

$$a = \frac{4A}{c\sin^2\theta} \qquad \text{...}(5)$$

A being the 'amplitude' of the concerned disturbance.

Now, θ being reasonably small, we can consider from fig. 1, the following figure,

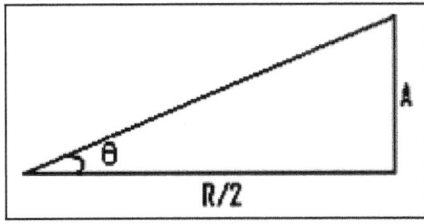

Fig. 4: Estimation of the disturbance due to spread of photon.

$$A = \frac{\lambda \tan \theta}{4}$$

$$= \frac{c \tan \theta}{4v} \qquad \qquad ...(6)$$

From equations (5) and (6), we get,

$$a = \frac{4A}{c \sin^2 \theta}$$

$$= \frac{1}{v \sin \theta \cos \theta} \qquad \qquad ...(7)$$

Finally, from equations (4) and (7), we have,

$$2cv \sin 2\theta = \frac{1}{v \sin \theta \cos \theta}$$

i.e.,

$$\theta = \frac{1}{2} \sin^{-1} \frac{1}{v\sqrt{c}} \qquad \qquad ...(8)$$

Minimum separation for two waves to be distinguishable:

From previous arguments it is obvious that there must be a certain minimum distance between two given disturbances, so that they are distinguishable. This minimum lateral distance ($\Delta \mathfrak{b}$) between the waves so that they can retain their individuality would depend on the difference in wavelengths ($\Delta \lambda$) for the two given waves.

In fact,

$$\Delta \mathfrak{b} \propto \Delta \lambda$$

Or,

$$\Delta \mathfrak{b} = \wedge \Delta \lambda \qquad \qquad ...(9);$$

where $\wedge$ is a constant of proportionality for the concerned system.

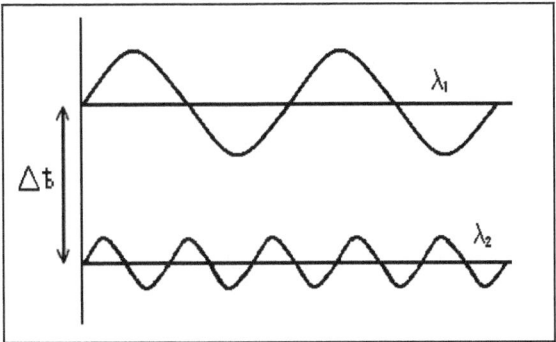

Fig. 5: Two waves of wavelengths λ_1 and λ_2 at a lateral distance $\Delta \bar{t}$ from each other. $\Delta \bar{t}$ is the minimum distance between the waves so that they are distinguishable.

We carry out an experiment to estimate the value of the constant

. The experimental set-up, as shown in Fig. 6, is used to pass a current through a cylindrical conductor M, through which electrons (due to skin effect) flow only along the surface. As such, an electric-current is obtained only along the surface of the conductor M. This conductor is again connected to a board T, in which conductors are layered one after another. The conductors in T and M are connected to electric bulbs. Inside M, an arrangement is made in such a way that a filament F sets out beams of photons of particular wavelengths. For various wavelengths we get current through various conductors, which can be seen from the glowing of various bulbs.

The photons, emanating from the filament with certain particular wavelengths force the electrons flowing along the skin of the cylindrical conductor M further away from it. We measure the wavelength of light (λ_L) for which the electrons move through a conductor, further away from the skin at a distance (say $\Delta \bar{t}$) from the filament.

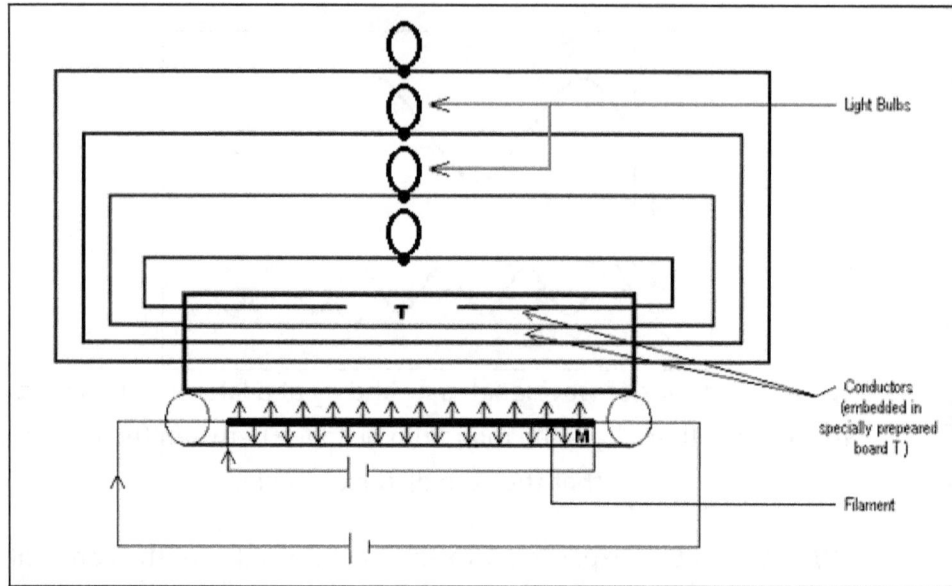

Fig. 6: Apparatus used to find out the value of λ The electrons have a kinetic energy,

$$K = 120eV$$

Therefore, momentum of electron,

$$p = \sqrt{2mK}$$

$$= \sqrt{2(9.11 \times 10^{-31})(120)(1.6 \times 10^{-19})}$$
$$= 5.91 \times 10^{-24} kg\ m/s$$

(the rest mass of electron $m = 9.11 \times 10^{-31} kg$)

Thus, de Broglie wavelength of electron,

$$\lambda = \frac{h}{p}$$

$$= \frac{6.63 \times 10^{-34} J.s}{5.91 \times 10^{-24} kg\ m/s}$$

$$= 112pm$$

Wavelength of light used,

$$\lambda_L = 560nm$$

The difference in wavelength of the light used and the electrons,
$\Delta\lambda = \lambda_L - \lambda_e$

$$= 560nm - 112pm$$
$$= 559.888nm$$
$$= 559.888 \times 10^{-9}m$$

The bulb that glows is due to the conductor at a distance (from the filament),
$\Delta b = 3.6 \times 10^{-4}m$

Therefore, the value of
$$\lambda = \frac{\Delta b}{\Delta\lambda} = \frac{3.6 \times 10^{-4}}{559.888 \times 10^{-9}}$$

i.e.,
$\lambda = 0.00642986 \times 10^{5}$

This experiment is repeated several times such that current is drawn through several other conductors embedded in the board T, such that the respective bulbs glow. In each case, Δb and $\Delta\lambda$ are noted   and   the   value   of

is calculated, which comes to quite the same in each case.

Inside the energy quantum:

In the lines of the treatment used in Chapter 1, on introduction of an energy quantum related to a spacetime configuration $[s' = (cT')^{4i}]$ in a region with configuration $\left[s = (cT)^{4i}\right]$, we may write,
$$hv = c^{4i}(\Delta T)^{4i} \qquad\qquad ... (10);$$
where, v is the frequency of the disturbance, h is the Plank's constant and $\Delta T = T' - T$.

Due to the introduction of the energy quantum, the vicinity gets 'stretched' from time T to T'. This stretching yields a cone as in

31

spacetime with a radius (say, r), height (ΔT) and slant-height (say, l).

The volume of this cone,

$$V = \frac{1}{3}\pi r^2 (\Delta T)$$

The increase in surface of the space-membrane,

$$\Delta \bar{s} = \pi r(l - r)$$

Thus, the ratio of the increase in the surface of the space-membrane to the 'volume' of the energy introduced,

$$n = \frac{\pi r(l - r)}{\frac{1}{3}\pi r^2 (\Delta T)}$$

i.e.,

$$n = \frac{3(l - r)}{r\Delta T}$$

Placing $l = \sqrt{r^2 + \Delta T^2}$ in the above equation, we get,

$$n = \frac{3\left(\sqrt{r^2 + \Delta T^2} - r\right)}{r\Delta T} \qquad \dots (11)$$

Now assuming the transport of the disturbed spacetime configuration to tend to take place equally in all directions as regards the laws of physical transactions, we may consider similar cones fill up all directions of spacetime. Thereby, we can say that the included angle of each such cone equals 60°.

Thus, for such a cone,

$$r = \Delta T \tan 30°$$

$$= \frac{\Delta T}{\sqrt{3}}$$

Placing this value of r in equation (11), we have,

$$n = \frac{3\left(\sqrt{\frac{\Delta T^2}{3} + \Delta T^2} - \frac{\Delta T}{\sqrt{3}}\right)}{\frac{\Delta T}{\sqrt{3}}\Delta T}$$

i.e.,

$$n = 3(2\Delta T - 1)$$

The increase in surface of the spacetime causes a wave in it. The frequency of the energy quantum-wave,

$$v = \frac{\text{increase in surface of spacetime}}{\Delta T}$$

$$= \frac{\pi r(l - r)}{\Delta T}$$

i.e.,

$$v = \pm \frac{\pi \Delta T}{3} \qquad \qquad ...(12)$$

The $'\pm'$ sign in equation (12) tallies with the existence of antiparticles to particles.

Now, placing $\Delta T = \pm \frac{3v}{\pi}$ from equation (12) in equation (10), we get,

$$hv = \pm \left(\frac{3cv}{\pi}\right)^{4i}$$

i.e.,

$$h = \pm \left(\frac{3c}{\pi}\right)^{4i} v^{4i-1} \qquad \qquad ...(13)$$

$\left(\frac{3c}{\pi}\right)^{4i}$ being considered constant,

$$h \propto v^{4i-1}$$

The value of h depends upon the frequency concerned. The graph of h versus v^{4i-1} (as shown in Fig. 7) is a monotonously increasing one. For most part of the energy-range that we deal with, h may be thus taken to be a constant. However, h seems not to be a constant in the strictest sense.

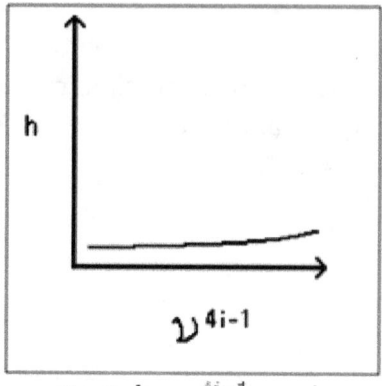

Fig. 7: $h - v^{4i-1}$ graph

Conclusion:

Though the angles of spread of various wavelengths are quite small, yet making use of an intermediate wavelength between two given wavelengths in an experimental set-up allows us to detect such spreading. Experimental values tally with theoretical ones.

A minimum distance must separate two given wavelengths for us to distinguish them. This minimum distance depends upon the difference between the concerned wavelengths. Experiment to study this gives us the value of the constant of proportionality between the two.

The energy quantum is studied in the light of the Theory of Abstraction and the laws governing physical transactions. The Plank's constant (h) appears to be a very monotonously increasing function instead of being a constant in the strictest sense of the term.

4 Abstraction of Observables

Key Ideas:

Making use of the laws of physical transactions, we study symmetrical many-points systems. Relation of group-theory to physical transactions in such symmetrical systems is dealt with. Studying perturbations in the stability states in the attractor-maps for transactions, approximate values of the observables are to be predicted for such systems. Further, Abstraction Theory is typified with respect to studying the properties of irreducible representations, if any, inside a given such group.

Introduction:

In previous chapters, the laws of physical transactions, in the light of Theory of Abstraction have been formulated. The mother equation (F= $\omega \frac{\lambda ST}{DR}$) describes the physical transactions taking place between two given points or between a given set of points in the vicinity of a concerned environment. A given point is influenced by its environment. On the other, it influences its concerned environment. A given point has some intrinsic properties. A group of such points form a field of extrinsic properties. The field of extrinsic properties, in turn, may influence the intrinsic characters of each of the individual points.

A set of points with same or of a similar-set of properties may be considered to belong to a given same group. For any given system, there can be one or a number of stability-states or symmetries. Further, each of such symmetries may have perturbations, affecting the average value of observable quantities. Measures of such perturbations are a useful way of finding approximate functions for systems when we know the exact transaction-functions for similar systems.

Average value of observables:

Let, f (λ, D) be a transaction function for a system, where λ is the difference in concentrations of a given observable quantity between two given points of transaction and D the distance between the points. Let, $f_1, f_2, f_3, ..., f_n$ be the complete orthonormal set of eigenfunctions for an operator Ô corresponding to some observable quantity in the system. f can be expanded such that,

$$f = j_1 f_1 + j_2 f_2 + j_3 f_3 + ... + j_n f_n \qquad ...(1)$$

where $j_1, j_2, j_3, ..., j_n$ are constants.

Operated with $\hat{O}$ from the left on both sides yield,

$$\hat{O}f = j_1\hat{O}f_1 + j_2\hat{O}f_2 + j_3\hat{O}f_3 + \ldots + j_n\hat{O}f_n$$

$f_1, f_2, f_3, \ldots, f_n$ being eigenfunctions of $\hat{O}$, we can write,

$$\hat{O}f = j_1 k_1 f_1 + j_2 k_2 f_2 + j_3 k_3 f_3 + \ldots + j_n k_n f_n$$

where $k_1, k_2, k_3, \ldots, k_n$ are the eigenvalues corresponding to the eigenfunctions.

Considering the complex conjugate of the transaction-function f in equation (1), we have,

$$f^* = j_1^* f_1^* + j_2^* f_2^*$$
$$+ j_3^* f_3^* + \ldots + j_n^* f_n^* \qquad \ldots (2)$$

Using these equations, after integrating over all co-ordinate space, we get,

$$\int f^* \hat{O}f \, dt$$

$$= j_1^* j_1 k_1 \int f_1^* f_1 \, dt$$

$$+ j_2^* j_2 k_2 \int f_2^* f_2 \, dt + \ldots + j_n^* j_n k_n \int f_n^* f_n \, dt \qquad \ldots (3)$$

In this equation, we have got rid of all the terms of the type $j_a^* j_b \hat{O}_b \int f_a^* f_b \, dt$, as these are all zero because of the orthogonality of the eigenfunctions. Only when $a = b$, are all the terms non-zero and these are the ones we have retained. The integrals on the right side of equation (3) are each equal to one because of the normality condition. Therefore, we write,

$$\int f^* \hat{O}f \, dt = j_1^* j_1 k_1 + j_2^* j_2 k_2 + \ldots + j_n^* j_n k_n$$

When the system is in the state f, the average value ($\bar{a}$) of the observable k is given by the right-hand side of the previous equation, such that,

$$\bar{a} = \int f^* \hat{O}f \, dt \qquad \ldots (4)$$

Relation of group theory to physical transactions in symmetrical systems:

Say a given dynamic system has a given set of symmetries or stability points. For all points having similar intrinsic properties within such a system, the probability densities of occurrence are equal and must remain unaltered, being all in a similar environment. Thus the energy and Hamiltonian for the system must not change. If E_i is the energy corresponding to the eigenfunction f_i, we may write,

$$\hat{H} f_i = E_i f_i$$

If a symmetry operation $(\hat{x})$ is performed on the system, we have,

$$\hat{x} \, \hat{H} f_i = \hat{x} \, E_i f_i$$

But since $\hat{x}$ does not affect $\hat{H}$ or E, we may write,

$$\hat{H} \, (\hat{x} f_i) = E_i \, (\hat{x} f_i)$$

The function $\hat{x} \, f_i$ is therefore an eigenfunction of $\hat{H}$ with the same eigenvalues as f_i. We can therefore conclude, if the state is non-degenerate, for normalized functions,

$$\hat{x} f_i = \pm f_i \qquad \qquad \ldots (5)$$

If a state (f_{in}) is degenerate with two or more eigenfunctions corresponding to a given energy, the energy can remain the same under the symmetry operation provided the original eigenfunction is transformed into a linear combination of the degenerate functions. For a $\Box\Box$-fold degenerate state,

$$\hat{x} f_{in} = \sum_{m=1}^{l} r_{mn} \, f_{in} \qquad \qquad \ldots (6)$$

where r_{mn} are the coefficients of the linear combination. They form representation matrices expressing the effect of the symmetry operations on the set of degenerate eigenfunctions f_{in} . The representation is irreducible.

This result is important as it relates the eigenfunctions of a system to its symmetry. It limits the forms of the eigenfunctions a symmetrical system can have.

Properties of Irreducible Representations of a Group:

The sum of the squares of the dimensions of the irreducible representations of a group is equal to the order of the group, in accordance with group-theory. The characters of the irreducible representations of a group behave as orthogonal vectors in an h-dimensional space; h being the order of the concerned group. If we label the character of the xth symmetry operation in the ith irreducible representation $y_i(x)$, this means that,

$$\sum_x y_i(x)\, y_j(x) = 0 \; ; \text{if } i \neq j \qquad \qquad \dots (7)$$

If we are dealing with complex characters, this equation would read

$$\sum_x [y_i(x)]^*\, y_j(x) = 0$$

If $i = j$, then,

$$\sum_x [y_i(x)]^2 = h \qquad \qquad \dots (8)$$

Combining equations (7) and (8), we get,

$$\sum_x y_i(x)\, y_j(x) = h\delta_{ij} \qquad \qquad \dots (9)$$

where δ_{ij} is the Kronecker delta.

Equation (9) expresses a necessary and sufficient condition that a representation is irreducible.

As the character of a given matrix remains the same after a similarity operation, for a particular similarity operation x, the sum of the characters of all the irreducible representations, we obtain from a reducible representation, is equal to the character of the reducible representation, such that,

$$y(x) = \sum_i a_i y_i(x)$$

where a_i is the number of times the ith irreducible representation occurs in the reducible representation.

Multiplying this equation for $y_i(x)$ and summing over all operations, we get,

$$\sum_x y(x)y_j(x) = \sum_i \sum_x a_i y_i(x)y_j(x) \qquad \ldots (10)$$

Substituting equation (9) into the right-side of equation (10), we have:

$$\sum_x y(x)y_j(x) = ha_j$$

a_j being the only remaining coefficient because the right-side of equation (10) is zero if $i \neq j$,

$$a_j = \frac{1}{h}\sum_x y(x)y_j(x) \qquad \ldots (11)$$

A reducible representation can be broken down into irreducible representations. Equation (11) gives a method of finding the number of times each irreducible representation occurs in a reducible representation.

Stability states in The Attractor-Maps For A Many-Points System:

Let a given system have (N) stability states or symmetries. (N) equals the number of types of intrinsic properties inside the system, i.e., N equals the number of groups inside the system. One possibility for the product-function for transactions may be written as,

$$f' = f_1(1)\, f_2(2)\, f_3(3), \ldots, f_n(N);$$

where $f_1, f_2, f_3, \ldots, f_n$ are the intrinsic transaction-functions for the groups 1, 2, 3,…,N, respectively.

As each of the stability states are otherwise indistinguishable (all being stability states), exchange of the co-ordinates of the stability states 1 and 2 amongst the intrinsic property groups will yield an equally good function,

$$f'' = f_1(2) \, f_2(1) \, f_3(3), \dots , f_N(N).$$

The number of functions of this type that can be written is N!, allowing for all possible exchanges of the co-ordinates of the stability states. The Slater determinant for an N-stability system is,

$$f = \frac{1}{\sqrt{N!}} \begin{vmatrix} f_1(1) & f_2(1) & \dots & f_N(1) \\ f_1(2) & f_2(2) & \dots & f_N(2) \\ \dots & \dots & \dots & \dots \\ f_1(N) & f_2(N) & \dots & f_n(N) \end{vmatrix} \qquad \dots (12)$$

where $\frac{1}{\sqrt{N!}}$ is a normalization factor.

Time Independent Perturbations:

Perturbation may be a useful way of predicting approximate functions for systems when we know the exact transaction-functions for similar systems. Let us consider a system for which we know the transaction-function f_i° and corresponding energy E_i°. These functions satisfy the equation,

$$\hat{H}^{\circ} f_i^{\circ} = E_i^{\circ} f_i^{\circ}$$

Let, there be a small perturbation that changes the functions to f_i; the energies change to E_i.

Let, the Hamiltonian for the perturbed system be Ĥ. We can then write,

$$\hat{H} f_i = E_i f_i \qquad \dots (13)$$

As the perturbation tends to zero, f_i tends to f_i°. We can therefore write,

$$f_i = f_i^{\circ} + \sum_{j \ne i} a_{ij} f_j^{\circ} \qquad \dots (14)$$

where a_{ij} are constants.

41

Substituting equations (14) into equation (13), and rearranging, we get,

$$\left(\hat{H} - E_i\right)f_i^\circ + \sum_{j \neq i} a_{ij}\left(\hat{H} - E_i\right)f_j^\circ = 0$$

We write the Hamiltonian as the sum of two parts, the unperturbed Hamiltonian $\hat{H}^\circ$ and a perturbation term $\hat{H}'$; i.e., $\hat{H} = \hat{H}^\circ + \hat{H}'$. Substituting this relation into the last equation, we get,

$$\left(\hat{H}^\circ + \hat{H}' - E_i\right)f_i^\circ + \sum_{j \neq i} a_{ij}\left(\hat{H}^\circ + \hat{H}' - E_i\right)f_j^\circ = 0$$

As, $\hat{H}^\circ f_i^\circ = E_i^\circ f_i^\circ$ and $\hat{H}^\circ f_j^\circ = E_j^\circ f_j^\circ$, the last equation becomes,

$$\left(E_i^\circ + \hat{H}' - E_i\right)f_i^\circ + \sum_{j \neq i} a_{ij}\left(E_j^\circ + \hat{H}' - E_i\right)f_j^\circ = 0$$

Multiplying by $f_i^{\circ*}$ from the left and integrating over all space, we get,

$$E_i^\circ \int f_i^{\circ*} f_i^\circ dt + \int f_i^{\circ*} \hat{H}' f_i^\circ dt - E_i \int f_i^{\circ*} f_i^\circ dt$$

$$+ \sum_{j \neq i} E_j^\circ a_{ij} \int f_i^{\circ*} f_j^\circ dt + \sum_{j \neq i} a_{ij} \int f_i^{\circ*} \hat{H}' f_j^\circ dt$$

$$- \sum_{j \neq i} E_i a_{ij} \int f_i^{\circ*} f_j^\circ dt = 0$$

Because of the orthonormality of the f_i°, we can write,

$$\int f_i^{\circ*} f_i^\circ dt = 1, \qquad \int f_i^{\circ*} f_j^\circ dt = 0,$$

which transforms the last equation into,

$$E_i^\circ - E_i + H_{ii}' + \sum_{j \neq i} a_{ij} H_{ij}' = 0 \qquad \qquad \text{... (15)}$$

where, $H_{ij}' = \int f_i^{\circ*} \hat{H}' f_j^\circ dt$

Multiplying from the left by $f_k^{'*}$; where $k \neq i$ and after manipulation, we get,

$$a_{ik}\left(E_k^o - E_i\right) + H_{ki}' + a_{ik}H_{kk}' + \sum_{j=i,k} a_{ij} H_{kj}' = 0$$

We now consider that the energies for the perturbed system, E_i and the coefficients a_{ij} can be written in terms of series in which successive terms become smaller, such that,

$$E_i = E_i^o + E_i' + E_i'' + \cdots$$
$$a_{ij} = a_{ij}' + a_{ij}'' + \cdots ;$$

where a single prime denotes a first-order term, a double prime denotes a second order term and so on.

Placing these last conditions into equation (15) and taking out the first-order terms (as two first-order terms multiplied together constitute a second-order term and so on), we get,

$$E_i' = H_{ij}' = \int f_i^{'*} \hat{H}' f_j^o \, dt \qquad \qquad \ldots (16)$$

Formation of Poles:

Each of the intrinsic properties contained inside a system will, for obvious reasons (in accordance with the Theory of Abstraction) transact in such a manner so as to be distributed as much as possible. This will give rise to two sets of transactions:

1. The transaction of the points themselves in every direction, including towards each other. This gives rise to an additional effect in the direction between two given points as compared to all the other directions. This is due to the fact that in the direction between the points, there is an effect due to both the points, while in all other directions (considering a two-point system) the effect is due to one point only as shown in fig.1

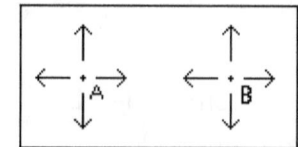

Fig. 1: effect between points

This gives rise to an 'attraction' between the points.

2. As a given intrinsic property transports in such a manner so as to be distributed as much as possible, the effect of 'repulsion' is generated. Let us consider two clusters of charges, either both positive or both negative. Not only the charge tends to be distributed in all directions, but also the type of charge. Presence of same type of charges will thus give rise to an effect of a field of repulsion in the vicinity of the clusters.

Thus we see that formation of poles and attraction between unlike poles and repulsion between like ones may be explained using the Theory of Abstraction. Here, we have examined only the qualitative aspects, while the quantitative aspects may be determined as detailed in my earlier chapter, in view of the laws of physical transactions.

Conclusion:

Abstraction Theory is made use of to study symmetrical many-points systems. The intrinsic properties of each of the constituents of the system need to be treated in accordance. We may relate each of the concerned points inside the given system to various groups, as per their intrinsic properties. A group of such points i.e., a system having individual intrinsic properties inside it may be regarded as a point with its own intrinsic properties (a function of its the constituent intrinsic properties) when it is considered as a part of some larger system. This way, intrinsic properties of each of the smallest points may influence a system or even a group of systems. A group of such individual points may give rise to a field of extrinsic properties, affecting each of the individual points.

Considering the symmetries or the stability states inside a given group or inside a given system as being similar in the fact that they are all basically states of stability, the irreducible representations, if any, inside a given system may be studied. Perturbations regarding a given average observable value may be predicted in a similar way.

5 Analysis of the Theory of Abstraction

Key Ideas:

In this chapter, a few more implications of the laws of physical transactions as per the Theory of Abstraction are dealt with. Analysis of those implications suggests the existence of 'hidden' mass and 'hidden' energy in a given physical transaction. Trajectory –examination of such possible transport is carried out. Relativistic cyclist phenomena are also dealt with in this chapter.

Introduction:

A particle in an isolated box will tend to move in all possible directions. A bias towards any given direction indicates an imbalance of support towards its movement in that given direction and resistance against it. Considering the movement of an energy quantum in a particular direction, this difference between the concerned support and the concerned resistance must be at least the same of the given energy quantum, in accordance to the Theory of Abstraction. For a given quantum-state ($h\upsilon$),

$$S \sim R = h\upsilon \qquad \qquad \dots (1)$$

where, S and R represent the support and resistance respectively.

This means that at least one half of a total energy-quantum gives it its direction while the other part gives it its magnitude. The direction part remains 'hidden' while only the magnitude part shows up as the value of the given quantum state. Considering the direction part however may reduce quantum-transport to classical transport as we shall see here.

Let us consider a photon revolving around a given body of mass (M) and within a radius (R). Let m_n be the instantaneous mass of the given photon. Then, according to the classical theory of gravitation, the force of gravity on the total photon (magnitude + direction parts) due to the given body equals

$$\frac{GM(m_n + m_n)}{R^2},$$

i.e.,

$$\frac{2GMm_n}{R^2}$$

In order that the photon does not move away from the body, the difference between support and resistance in that given direction must be zero. Thus, in order to bind the photon within a radius R, only its direction-part needs to be bound.

Now the outward force experienced by the direction-part of the photon equals $\frac{m_n c^2}{R}$; vaccum being the medium considered.

For light to be bound within the radius R,

$$\frac{m_n c^2}{R} - \frac{2GMm_n}{R^2} \leq 0$$

Choosing the equality sign,

$$m_n \left(c^2 - \frac{2GM}{R} \right) = 0$$

For the energy-quantum to exist,

$$m_n \neq 0$$

Therefore,

$$c^2 - \frac{2GM}{R} = 0$$

i.e.,

$$\frac{M}{R} = \frac{c^2}{2G} \approx 10^{27}\,kg/m \qquad\qquad \ldots(2)$$

Thus, if mass is concentrated within a radius R such that $\frac{M}{R} \approx 10^{27}\,kg/m$, then the body is a black-hole, in accordance with the Theory of Abstraction. This result is validated by existing data.

In this treatment, considering the direction-part of a given energy-quantum, classical transport merges with quantum-transport quite perfectly. Support towards the given direction of transport of a given energy-quantum comprises the direction-part of it while a given resistance may act upon the total energy-quantum i.e., the direction-part and the magnitude-part combined.

Chaos-Analysis of Quantum-Transport:

For a given transport of energy-quantum, between an initial and a final point, let the trajectory of the initial point $x_o = x(o)$ be denoted by,

$$x(t) = f^t(x_o)$$

Expanding $f^t(x_o + \delta x_o)$ to linear order, the evolution of the distance to a neighbouring trajectory $x_i(t) + \delta x_i(t)$ is given by the Jacobian matrix J,

$$\delta x_i(t) = \sum_{j=1}^{d} J^t(x_o)_{ij}\ \delta x_{oj},$$

$$J^t(x_o)_{ij} = \frac{\delta x_i(t)}{\delta x_{oj}} \qquad\qquad \ldots(3)$$

A trajectory of an energy-quantum as moving on a flat surface is specified by two position coordinates and the direction of motion. The Jacobian matrix describes the deformation of an infinitesimal neighbourhood of $x(t)$ along the transport. Its eigenvectors and eigenvalues give the directions and the corresponding rates of expansion or contraction. The trajectories that start out in an infinitesimal neighbourhood separate along the unstable directions

(those whose eigenvalues are greater than unity in magnitude), approach each other along the stable direc-
tions (those whose eigenvalues are less than unity in magnitude), and maintain their distance along the marginal directions (those whose eigenvalues equal unity in magnitude).

Holding the hyperbolicity assumption (i.e., for large n the prefactors a_i, reflecting the overall size of the system, are overwhelmed by the exponential growth of the unstable eigenvalues Λ_i, and may thus be neglected), to be justified, we may replace the magnitude of the area of the ith strip $|B_i|$ by $\frac{1}{|\Lambda_i|}$ and consider the sum,

$$\lceil n = \sum_i^n \frac{1}{|\Lambda_i|};$$

where the sum goes over all periodic points of period n. We now define a generating function for sums over all periodic orbits of all lengths,

$$\lceil z = \sum_{n=1}^{\infty} \lceil n\, z^n \qquad\qquad \dots (4)$$

For large n, the nth level sum tends to the limit $\lceil n \to e^{-n\gamma}$, so the escape rate γ is determined by the smallest $z = e^\gamma$ for which equation (4) diverges,

$$\lceil z \approx \sum_{n=1}^{\infty} (ze^{-\gamma})^n = \frac{ze^{-\gamma}}{1 - ze^{-\gamma}} \qquad\qquad \dots (5)$$

Making an analogy to the Riemann zeta-function, for periodic orbit cycles,

$$\lceil z = -z\frac{d}{dx} \sum_p \ln\left(1 - t_p\right);$$

$\lceil(z)$ is a logarithmic derivative of the infinite product

$$\frac{1}{\zeta(z)} = \prod_p (1 - t_p), \, t_p = \frac{z^{n_p}}{|\Lambda_p|}$$

This represents the dynamical zeta function for the escape rate of the trajectories of quantum-transport. The fraction of initial x whose trajectories remain within B at time t may decay exponentially,

$$\lceil t = \frac{\int_s dx \, dy \, \delta[y - f^t(x)]}{\int_s dx} \rightarrow e^{-\gamma t}.$$

The integral over x starts a trajectory at every $x \, \epsilon \, B$. The integral over y tests if this trajectory still falls within limits of B at time t.

The Kernel of this integral is the evolution operator for a d-dimensional transport,

$$\mathcal{L}^t(y, x) = \delta[y - f^t(x)] \qquad \qquad \dots (6)$$

Expressing the finite time Kernel $\mathcal{L}^t$ in terms of A, the generator of infinite time translations,

$$\mathcal{L}^t = e^{tA} \qquad \qquad \dots (7)$$

This is very much similar to the way quantum evolution is generated by the Hamiltonian.

Relativity of Cyclists and Symmetry Analysis:

The component of the dynamics along the continuous symmetry directions of the trajectory behavior or 'drift' may be induced by the symmetries themselves. In presence of a continuous symmetry, an orbit explores the manifold swept by combined actions of the dynamics and the symmetry induced drifts. A group member can be parameterized by angle θ, with the group multiplication law $g(\theta') \, g(\theta) = g(\theta' + \theta)$ and its action on smooth periodic functions $u(\theta + 2\Pi) = u(\theta)$ generated by,

$$g(\theta') = e^{\theta' T}, T = \frac{d}{d\theta}$$

The differential operator T is reminiscent of the generator of spatial translations. The constant velocity field $v(x) = v = C.T$ acts on x, by replacing it by the velocity vector C_j.

Let, G be a group and $gB \to B$ a group action on the state space B. The $[d \times d]$ matrices g acting on vectors in the d-dimensional state space B from a linear representation of the group G. If the action of every element g of a group G commutes with the flow,

$$g\, v(x) = v(gx), gf^t(x) = f^t(gx)$$

G is a symmetry of the dynamics and is invariant under G or G-equivalent. For any $x \in B$, the group orbit B_x of x is the set of all group actions,

$$B_x = [g\, x \mid g \in G] \qquad \qquad \qquad ...(8)$$

The time evolution and the continuous symmetries can be considered on the same Lie group footing. An element of a compact Lie group continuously connected to identity can be written as,

$$g(\theta) = e^{\theta T}, \theta.T = \sum \theta_a T_a, a = 1,2,3,...,N;$$

where $\theta.T$ is a Lie element and θ_a are the parameters of the transformation.

Any representation of a compact Lie group G is fully reducible, and invariant tensors constructed by contractions of T_a are useful for identifying irreducible representations. The simplest such invariant is,

$$T^T.T = \sum_x c_2^{(\alpha)} \|^{(\alpha)}$$

equilibria satisfy $f^t(x) - x = 0$ and relative equilibria satisfy $f^t(x) - g(t)x = 0$ for any t. A relative periodic orbit is periodic in its mean velocity, $C_p = \theta_p/T_p$, comoving frame, but in the stationary frame its trajectory is quasiperiodic. A relative periodic orbit may be pre-periodic if it is equivariant under a discrete symmetry.

Subhajit Ganguly

Translational symmetry allows for relative equilibria characterized by a fixed profile Eulerian velocity $\mu_{TW}(x)$ moving with constant velocity C, i.e.,

$$\mu(x,t) = \mu_{TW}(x - Ct) \qquad \ldots (9)$$

A relative periodic solution is a solution that recurs at time T_p with exactly the same disposition of the Eulerian velocity fields over the entire cell, but shifted by a 2-dimensional (streamwise–spanwise) translational g_y. By discrete symmetries, these solutions come in counter-travelling pairs,

$$u_q(x - Ct), -u_q(-x + Ct).$$

The Hilbert basis may be written as,

$$u_1 = x_1^2 + x_2^2, u_2 = y_1^2 + y_2^2, u_3 = x_1 y_2 - x_2 y_1, u_4$$
$$= x_1 y_1 + x_2 y_2, u_5 = z$$

This is invariant under the special orthogonal graph SO(2);i.e., a group of length-preserving rotations in a plane. The polynomials are linearly independent, but related through one syzygy,

$$u_1 u_2 - u_3^2 - u_4^2 = 0 \qquad \ldots (10)$$

yielding a 4-dimensional B/ SO(2) reduced state-space.

The dynamical equations following from chain rule are,

$$\dot{u}_i = \frac{\delta u_i}{\delta x_j} x_j \qquad \ldots (11).$$

Conclusion:

From the Theory of Abstraction, we arrive at 'hidden' direction part of an energy-quantum. Quantum dynamics is seen to merge with classical dynamics if this hidden direction-part of the quantum-states are taken into consideration, as validated by practical analysis and data. Moreover, this hidden part of an actual energy-quantum may explain the dark-energy problem. As a support towards transport comprises the direction-part only, and as the resistance against motion is offered against the whole of an energy-quantum (direction-part + magnitude-part), this hidden energy may very well affect a gravitational field. On the other hand, if matter is considered to be condensed energy, and as such condensed energy quanta in some form of orientation in space time, there ought to be some self-same hidden mass as the magnitude-parts of the constituent energy quanta, thereby indicating a hidden dark-matter. Thus the anomaly of the existence of dark-matter and dark-energy being hidden from us may be reconciled with in accordance with the Theory of Abstraction.

6 Hamiltonian Dynamics in the Theory of Abstraction

Key Ideas:

This chapter deals with fluid flow dynamics which may be Hamiltonian in nature and yet chaotic. Here we deal with sympletic invariance, canonical transformations and stability of such Hamiltonian flows. As a collection of points move along, it carries along and distorts its own neighbourhood. This in turn affects the stability of such flows.

Introduction:

There may be many settings of physical interest, where dynamics is reversible, such as finite-dimensional Hamiltonian flows. In such transactions, the family of evolution maps f^t form a group. The evolution rule f^t is a family of mappings of strips of transport B, that we may consider, such that,

1) $f^0(x) = x$
2) $f^t[f^{t'}(x)] = f^{t+t'}(x)$

3) $(x, t) \rightarrow f^t(x)$ from $B \times R$ into B is continuous; where t represents a time interval and $t \in R$.

For infinitesimal times, we may write the trajectory of a given transaction as,

$$x(t + \tau) = f^{t+\tau}(x_0)$$
$$= f[f(x_0, t), \tau]$$

The time derivative of this trajectory at point $x(t)$ is,

$$\frac{dx}{d\tau}\bigg|_{\tau=0} = \partial_\tau f[f(x_0, t), \tau]|_{\tau=0} = \dot{x}(t)$$

The vector field is a generalized velocity field,

$$\dot{x}(t) = v(x) \qquad \qquad \text{...(1)}$$

If x_q represents an equilibrium point, the trajectory remains stuck at x_q forever. Otherwise, the trajectory passing through x_0 at time $t = 0$ may be obtained by,

$$x(t) = f^t(x_0) = x_0 + \int_0^t d\tau \, v[x(\tau)], x(0)$$

$$= x_0 \qquad \qquad \text{...(2)}$$

The Euler integrator, which advances the trajectory by $\delta\tau \times$ velocity at each time step is,

$$x_i = x_i + v_i(x)\delta\tau.$$

This may be used to integrate the equations of the dynamics concerned.

Hamiltonian Chaotic Dynamics:

For a Hamiltonian $H(q, p)$ and equations of motion

$$\dot{q}_i = \frac{\partial H}{\partial p_i}, \dot{p}_i = \frac{\partial H}{\partial q_i}$$

With D degrees of freedom,

$x = (q, p)$,

$q = (q_1, q_2, q_3, \ldots, q_D)$,

$p = (p_1, p_2, p_3, \ldots, p_D)$.

The value of the Hamiltonian function at the state space point $x = (q, p)$ is constant along the trajectory $x(t)$. Thus the energy along the trajectory $x(t)$ is constant,

$$\frac{d}{dt} H[q(t), p(t)] = \frac{\partial H}{\partial q_i} \dot{q}_i(t) + \frac{\partial H}{\partial p_i} \dot{p}_i(t) = \frac{\partial H}{\partial q_i} \frac{\partial H}{\partial p_i} - \frac{\partial H}{\partial p_i} \frac{\partial H}{\partial q_i} = 0$$

The trajectories therefore lie on surfaces of constant energy or level sets of the Hamiltonian $[(q, p): H(q, p) = E]$.

Given a smooth function $g(x)$, the standard map is,

$x_{n+1} = x_n + y_{n+1}$

$y_{n+1} = y_n + g(x_n)$.

This is an area-preserving map. The corresponding n^{th} iterate Jacobian matrix is,

$$M^n(x_0, y_0) = \prod_{K=n}^{1} \begin{pmatrix} 1 + g'(x_K) & 1 \\ g'(x_K) & 1 \end{pmatrix}$$

Let $M = 1$ as the map preserves areas and also B is symplectic. The standard map corresponds to the choice $g(x) = k/(2\pi \sin(2\pi x))$. When $k = 0$, $y_{n+1} = y_n = y_0$, so that angular momentum is conserved, and the angle x rotates with uniform velocity,

$$x_{n+1} = x_n + y_0 = x_0 + (n+1)y_0 \qquad \ldots (3)$$

The standard map provides a stroboscopic view of the flow generated by a time dependent Hamiltonian,

$$H(x, y, t) = \frac{1}{2} y^2 + G(x)\delta_1(t);$$

where δ_1 denotes the periodic delta function,

$$\delta_1(t) = \sum_{m=-\infty}^{\infty} \delta(t - m)$$

and

$$G'(x) = -g(x) \qquad\qquad \text{... (4)}$$

A complete description of the dynamics for arbitrary values of the nonlinear parameter k is fairly complex. When K is sufficiently large, single trajectories wander erratically on a large fraction of the phase space.

Stability In Hamiltonian Flows:

The equations of motion for a time-independent D-degrees of freedom Hamiltonian can be written as,

$$\dot{x}_i = \omega_{ij}H_j(x), \omega = \begin{pmatrix} 0 & -I \\ -I & 0 \end{pmatrix}, H_j(x) = \frac{\partial}{\partial x_j}H(x);$$

where $x = (q, p) \in B$ is a phase space point. $H_K = \partial_K H$ is the column vector of partial derivatives of H, I is the $[D \times D]$ unit matrix and ω the $[2D \times 2D]$ symplectic form.

$$\omega^T = -\omega, \omega^2 = -1$$

The evolution of J^t is determined by the stability matrix A,

$$\frac{d}{dt}J^t(x) = A(x)J^t(x),$$

$$A_{ij}(x) = \omega_{ik}H_{kj}(x);$$

where the matrix of second derivations $H_{kn} = \partial_k\partial_n H$ is the Hessian matrix. For symmetry of H_{kn}, $A^T\omega + \omega A = 0$.

This is the defining property for infinitesimal generators of symplectic or canonical transformations, which leave the symplectic form ω invariant.

For an equilibrium point x_q the stability matrix A is constant. The characteristic polynomial of A for an equilibrium x_q satisfies,

$$\det(A - \lambda I) = \det\left[\omega^{-1}(A - \lambda I)\omega\right]$$
$$= \det[-\omega A\omega - \lambda I]$$
$$= \det(A^T + \lambda I) = \det(A + \lambda I)$$

The symplectic invariance implies that if λ is an eigenvalues, then $-\lambda, \lambda^*$ and $-\lambda^*$ are also eigenvalues.

A Floquet multiplier $\Lambda = \Lambda(x_0, t)$ associated to a trajectory is an eigenvalues of the Jacobian matrix J and it satisfies

$$\det(J - \Lambda I) = \det(J^T - \Lambda I) = \det(-\omega J^T \omega - \Lambda I)$$
$$= \det(J^{-1}) \det(I - \Lambda J)$$
$$= \Lambda^{2D} \det(J - \Lambda^{-1} I) \qquad \dots (5)$$

This is because, $J^{-1} = -\omega J^T \omega$, J being symplectic. If Λ is an eigenvalue of J so are $\frac{1}{\Lambda}, \Lambda^*$ and $\frac{1}{\Lambda^*}$. Real eigenvalues always come paired as $\Lambda, \frac{1}{\Lambda}$. The complex eigenvalues come in pairs $\Lambda, \Lambda^*, |\Lambda| = 1$, or in loxodromic quartets $\Lambda, \frac{1}{\Lambda}, \Lambda^*$ and $\frac{1}{\Lambda^*}$.

For a trajectory originating near $x_0 = x(0)$ with an initial infinitesimal displacement $\delta x(0)$, the flow transports the displacement $\delta x(t)$ along the trajectory $x(x_0, t) = f^t(x_0)$.

This infinitesimal displacement is transported along the trajectory $x(x_0, t)$, with time variation given by,

$$\frac{d}{dt} \delta x_i(x_0, t)$$

$$= \sum_j \left. \frac{\partial v_i}{\partial x_j}(x) \right|_{x=x(x_0,t)} \delta x_j(x_0, t) \qquad \dots (6)$$

The system of linear equations of variations for the displacement of the infinitesimally close neighbour $x + \delta x$ follows from the flow equations by Taylor expansion to linear order,

$$\dot{x}_i + \dot{\delta x}_i = v_i(x + \delta x)$$

$$\approx v_i(x) + \sum_j \frac{\partial v_i}{\partial x_j} \delta x_j \qquad \dots (7)$$

Taking these together, the set of equations governing the dynamics in the tangent bundle $(x, \delta x) \in TB$ obtained by adjoining the d-dimensional tangent space $\delta x \in TB_x$ to every point $x \in B$ in the d-dimensional state space $B \subset R^D$, may be written as,

$$\dot{x}_i = v_i(x), \dot{\delta x}_i = \sum_j A_{ij}(x) \delta x_j$$

and the stability matrix or the velocity gradients matrix may be written as,

$$A_{ij}(x) = \frac{\partial v_i(x)}{\partial x_j} \qquad \ldots (8)$$

Equation (8) describes the instantaneous rate of shearing of the infinitesimal neighbourhood of $x(t)$ by the flow.

Description of neighbourhoods of equilibria and periodic orbits is afforded by projection operators,

$$P_i = \prod_{j=i} \frac{M - \lambda^{(j)} I}{\lambda^{(i)} - \lambda^{(j)}} \qquad \ldots (9);$$

where matrix M is either equilibrium stability matrix A, or periodic Jacobian matrix $\hat{J}$ restricted to a Poincare section. For each distinct eigenvalue $\lambda^{(i)}$ of M, the columns/rows of P_i are,

$$(M - \lambda^{(i)})P_j = P_j(M - \lambda^{(j)} I) = 0$$

are the right/left eigenvectors $e^{(K)}, e_{(K)}$ of M which span the corresponding linearized subspace, provided M is not of Jordan type. For determining the eigenvalues of the finite time local deformation J^t for a general nonlinear smooth flow, the Jacobian matrix may be computed by integrating the equations of variations,

$$x(t) = f^t(x_0), \delta x(x_0, t) = J^t(x_0)\delta x(x_0, 0)$$

For equilibrium point x_q,

$$J^T(x_q) = e^{A_q t}, A_q = A(x_q) \qquad \ldots (10)$$

Conclusion:

In Hamiltonian dynamics of real physical entities, the difference between support towards a flow and resistance against it not only causes the transactions to take place, but also distorts the neighbourhood of the flow. For sufficiently large values of K, a given Hamiltonian flow may turn into a highly chaotic one. In accordance with the Laws of Physical Transactions as per the

Theory of Abstraction, such flows will generate disturbances in all parts of the vicinity of the transactions. These disturbances will tend to smooth out equally in every possible direction, causing secondary disturbances against stability.

7 Condensation States and Landscaping with the Theory of Abstraction

Key Ideas:

The Abstraction Theory is applied in landscaping. A collection of objects may be made to be vast or meager depending upon the scale of observations. This idea may be developed to unite the worlds of the great vastness of the universe and the minuteness of the sub-atomic realm. Keeping constant a scaling ratio for both worlds, these may actually be converted into two self-same representatives with respect to scaling. The Laws of Physical Transactions are made use of to study Bose-Einstein condensation. As the packing density of concerned constituents increase to a certain critical value, there may be evolution of energy from the system.

Introduction:

Be it the large vastness of the universe or the delicate smallness of the sub-atomic world, by choosing a suitable constant scaling ratio for both, we may obtain their representations. These representations, following a certain constant scaling ratio, will be self-same. In the previous chapters, I have mentioned the chaotic behavior in the quantum world. Choosing suitable scaling ratios, we may turn the universe itself into such a chaotic quantum system, having its own necessary quantum states and trajectory behaviour. In that case, the study of the universe reduces to the study of some sort of a quantum chaotic system. On the other hand, choosing some other necessary scaling ratios, the atomic and the sub-atomic realm may be extended to become the universe itself, complete with its own macroscopic trajectory behaviour. Instead of formulating different ways of looking at worlds of different sizes, if we adjust the way of viewing i.e., the scaling ratios in such a fashion that the representations of the worlds merge, we will be looking at representative worlds of study which are practically self-same. The Laws of Physical Transactions formulated in previous chapters may then be applied in order to study such self-same representations of the worlds of various scales. Unification of the ways of studying at different ranges of scaling may thus be achieved by suitable landscaping (adjusting different scales to a suitable scaling ratio, in order to make all the scales of study similar in size). Further, a similar approach may be applied to study the Bose-Einstein Condensation. A certain critical packing density of the constituents of each world of a certain landscape must ensure a condensation of similar sort. The quantum states (or some similar states) of each such landscape will merge and give spikes for that critical scaling ratio in their respective representations.

The quantum chaotic behavior may be of interest to study if we are to learn about the universe as a whole. The astronomically large distances separating clusters in the universe supports a study of such sorts. Quantum chaotic behaviour, on the other hand, will give rise to something similar to the Bose-Einstein Condensation at some critical packing density. The study of such condensation states too will be of interest here.

Scaling The Universe:

Looking at a large enough part of the universe, we may draw an analogy to a system of scattered particles, in motion or rest, relative to each other. These particles may or may not be similar to each other, if we look at a given locality. Our idea, however, is that we can always represent even the whole of the universe on a piece of paper of our desired size. We can very well do the same with localities of sub-atomic sizes.

We may represent both the worlds, viz., the microscopic and the macroscopic, within any desired standard size. Theoretically, we are only to diminish the snaps of the universe and magnify the snaps of the microscopic world in order to put both into representations of a definite scaling-size. Looking at such a representation of the macroscopic world (due to the large number of constituents and the large distances separating them involved) we will find it to be a complex mixture of various kinds of particles. On the other hand, looking at such a representation of the microscopic world, (due to the small distances separating the constituents) it will be like the actual universe itself, with various types of constituent parts involved. Such a representation of the microscopic and the macroscopic worlds will bring out hidden properties and behaviours of both worlds, as well as providing for a similar basis of studying them both.

Let us consider a given representation with fractal dimension D_F. The fractal dimension is purely geometrical, i.e., it only depends on the shape of the representation. A suitable probability measure $d\mu$, according to the particular phenomenon considered is assigned to the given representation. A coarse grained probability density, as the mass of the hypercube Λ_i of size l is defined as,

$$P_i(l) = \int_{\Lambda_i} d\mu(x) \qquad \qquad \dots (1);$$

where $i = 1,2,3,\dots,N(l)$.

The information dimension D_I is such that,

$$\sum_{i=1}^{N(l)} P_i \ln(P_i) \simeq D_I \ln(l) \qquad \qquad \dots (2);$$

where $D_I \leq D_F$.

The number of boxes containing the dominant contributions to the total mass and thus relevant part of the information, is,

$$N_R(l) \propto l^{-D_I} \qquad \qquad \dots (3).$$

For each box Λ_i, $D_I = D_F$ for a uniform distribution. When $D_I < D_F$, the measure itself may be called fractal since it is singular with respect to the uniform distribution,

$$P^* = \frac{1}{N(l)} \propto l^{D_F}$$

for each box Λ_i. Thus, $\frac{P_i}{P_i^*}$ can diverge in the limit of vanishing l.

Simulations of the mass-moment scaling yields,

$$\langle P_i(l)^q \rangle \equiv \sum_{i=1}^{N(l)} P_i(l)^{q+1} \propto l^{q \cdot d_q + 1} \qquad \qquad \dots (4).$$

The d_q are the Renyi dimensions which generalize the information dimension $D_I = d_1$ as well as the fractal dimension $D_F = d_0$. If the d_q's are not constant, anomalous scaling is to be employed and, as the order q varies, the amount of the difference $D_q - D_F$ gives a first rough measure of the heterogeneity of the probability distribution.

The moment generic observables A computed on scale l is such that,

$$\langle A(l)^q \rangle \propto l^{g(q)}$$

Anomalous scaling, i.e., a non-linear shape of the function $g(q)$ is the more common situation, where one does not require unnecessarily to consider only a finite number of scaling components. In some cases, one may observe strong time variations in the degree of chaoticity. This intermittency phenomenon involves an anomalous scaling with respect to time-dialations identifying the parameter e^{-t} with the parameter l used in spatial dialations. A measure of the degree of intermittency requires the introduction of infinite sets of exponents which are analogous to the Renyi dimensions and can be related to a multifractal structure given by the dynamical system in the functional trajectory space.

The Grassberger-Procaccia correlation dimension v is defined by considering the scaling of the correlation integral,

$$C(l) = \lim_{M \to \infty} \frac{1}{M^2} \sum_i \sum_{j \neq i} \theta \left(l - |x_i - x_j| \right);$$

where θ is the Heaviside step function and $C(l)$ is the percentage of pairs (x_i, x_j) with distance $|x_i - x_j| \leq l$.

In the limit $l \to 0$,

$$C(l) \propto l^v.$$

In general,

$$v \leq D_F.$$

v is a more relevant scaling index than D_F since it is related to the point probability distribution on the attractor, while D_F cannot take into account an eventual homogeneity in the visit frequencies.

Let us define the number of points in an F-dimensional spherical representation of the world, with radius l and centre at x_i as,

$$n_i(l) = \lim_{M \to \infty} \frac{1}{M-1} \sum_{j \neq i} \theta \left(l - |x_i - x_j| \right) \qquad \ldots (5).$$

We must introduce a whole set of generalized scaling exponents

$$\langle n(l)^q \rangle = \lim_{M \to \infty} \frac{1}{M} \sum_{i=1}^{M} n_i(l)^q \propto l^{\emptyset(q)} \qquad \ldots (6);$$

where $\emptyset(1) = v$.

Considering a uniform partition of phase space into boxes of size l it is convenient to introduce the probability $P_K(l)$ that a point x_i falls into the Kth box. In this case, the moments of P_K can be estimated by summing up the boxes,

$$\langle p(l)^q \rangle = \sum_{K=1}^{N(l)} P_K(l)^{q+1} \propto l^{q \cdot d_{q+1}},$$

A moment of reflection shows

$$\emptyset(q)/q = d_{q+1}$$

because of the ergodicity $n_i(l) \sim P_K(l)$, if x_i belongs to the Kth box and since one can use either an 'ensemble' average (weighted sum over the boxes) or a 'temporal' average (sum of the time evolution $x(l)$).

The fractal dimension for $q = -1$ is,

$$D_F = d_0 = -\emptyset(-1) \qquad \qquad \text{... (7)}$$

while the correlation dimension is,

$$v = d_2 = \emptyset(1) \qquad \qquad \text{... (8)}$$

Statistical laws at small scales have to depend not only on the average energy dissipation density $\bar{\varepsilon}$ but also on the fluctuations of energy dissipation density $\varepsilon(x)$.

According to the Theory of Physical Abstraction, each point x should have the same singularity structure,

$$\Delta V_x(r) \propto r^h, h = \frac{1}{3} \qquad \qquad \text{... (9)}$$

In other words $\varepsilon(x)$ tends to be smoothly distributed in a region of R^3. The eddy turn-over time and the kinetic energy per unit mass at scale r are defined as,

$$t(r) \sim \frac{r}{\Delta V(r)} \qquad \qquad \text{... (10)}$$

and

$$E(r) \sim \Delta V(r)^2 \qquad \qquad \text{... (11)}$$

The transfer rate of energy per unit mass from the eddy at scale r to smaller eddies is then given by

$$\tilde{\varepsilon}(r) = \frac{E(r)}{t(r)} \sim \frac{\Delta V(r)^3}{r} \qquad \qquad \text{... (12)}$$

Since

$$\varepsilon_x(r) = \left(\frac{1}{r^3}\right) \int_{\Lambda_x(r)} \varepsilon(y)d^3y,$$

$[\Lambda_x(r)$ is a cube of edge r around $x]$ we have,

$$\int_{\Lambda_x(r)} \varepsilon(y)d^3y \sim r^3 \qquad\qquad \dots (13)$$

$r \to 0$ means r in the initial range and the regions containing a large part of $\varepsilon(x)$ are a physical approximation of a fractal structure. In this β −model approach,

$$\int_{\Lambda_x(r)} \varepsilon(y)d^3y \propto \begin{cases} r^{D_F} & if\ x \in S \\ 0 & if\ x \notin S \end{cases}$$

in an equivalent way

$$\Delta V_x(r) \propto \begin{cases} r^h & if\ x \in S \\ 0 & if\ x \notin S; \end{cases}$$

where $h = (D_F - 2)/3$

At scale r,there is only a fraction,

$$r^{3-D_F} \propto \frac{r^{-D_F}}{r^{-3}}$$

occupied by active eddies.

The transfer energy from the eddy at scale l_n (active eddy) to the scale l_{n+1} is $\varepsilon_n \propto \frac{v_n^3}{l_n}$.

Since, the energy transfer rate is constant in the cascade process, for $\beta = 2^{D_F-3}$, we have,

$$\varepsilon_n = \beta \varepsilon_{n+1}, \frac{v_n^3}{l_n} = \beta \frac{v_{n+1}^3}{l_{n+1}} \qquad\qquad \dots(14)$$

Iterating, we have,

$$v_n \propto l_n^{1/3}(l_n/l_0)^{\frac{D_F-3}{3}} \qquad\qquad \dots(15)$$

Each eddy at scale l_n is divided into eddies of scale l_{n+1} in such a way that the energy transfer for a fraction β of eddies increases by a factor $\frac{1}{\beta}$, while it becomes zero for the other ones.

In order to generalize the β-model, we have at scale l_n, N_n active eddies. Each eddy $l_n(k)$ generates active eddies covering a fraction of volume $\beta_{n+1}(k)$. k labels the mother-eddy and $k = 1, \dots, N_n$.

Since the rate of energy transfer is constant among mother-eddies and their effects, we have,

$$\frac{v_n(k)^3}{l_n} = \beta_{n+1}(k) \frac{v_{n+1}(k)^3}{l_{n+1}} \qquad \dots (16)$$

The iteration of v_n gives an eddy generated by a particular history of fragmentations $[\beta_1, \dots, \beta_n]$, such that,

$$v_n \propto l_n^{1/3} \left(\prod_{i=1}^{n} \beta_i \right)^{-1/3} \qquad \dots (17)$$

The fraction of volume occupied by an eddy generated by $[\beta_1, \dots, \beta_n]$ is $\prod_{i=1}^{n} \beta_i$, such that,

$$\langle |\Delta V(l_n)|^P \rangle$$

$$\propto l_n^{P/3} \int \prod_{i=1}^{n} d\beta_i \; \beta_i^{(1-P/3)} P(\beta_1, \dots, \beta_n) \qquad \dots (18)$$

With no correlation among different steps of the fragmentation, i.e., with $P(\beta_1, \dots, \beta_n) = \prod_{i=1}^{n} P(\beta_i)$, the exponent concerned,

$$\zeta_P = \frac{P}{3} - \ln_2 \{ \beta^{(1-P/3)} \} \qquad \dots (19)$$

Let us now consider a given representation of the universe. Let the packing density of the constituents be Ø. This packing density function Ø will affect any given constituent point inside it in accordance with the Laws of Physical Transactions. The given constituent point concerned will in turn affect Ø while interacting. For a given critical state of study of the total effects, we therefore are going to have a shear stress Ø and a mean effective stress f.

The critical state line is the loci of critical state conditions in the $\varepsilon - f - \emptyset$ space. Its projection on the $f - \emptyset$ space defines a strength parameter,

$$M = \frac{\emptyset}{f} = \frac{6 \sin \emptyset}{3 - \sin \emptyset} \qquad \ldots (20).$$

The second equality applies to axis-symmetric, axial compression and it is a function of the constant volume critical state packing density function $\emptyset$.

The small-strain stiffness of a given representation is measured by imposing a smaller strain than the elastic threshold strain concerned. In this range, deformations localize at inter-point contacts and the granular skeleton deforms at constant fabric of spacetime. The nonlinear load-deformation response determines the stress-dependent shear wave velocity,

$$V_s \propto \left(\frac{f - \emptyset}{\varepsilon}\right)^\beta \qquad \ldots (21)$$

Inside a given cluster, we may have various growth-patterns. The growth may occur mainly at an active zone on the surface of the cluster. For a one-dimensional interface, a fluctuation-dissipation theorem exists, leading to an exact dynamic exponent $z = \frac{3}{2}$. This is in excellent agreement with numerical simulations of ballistic aggregation and Eden clusters. For two-dimensional interfaces, $z \sim 1.5$.

The interface profile is described by a height $h(x, t)$. The simplest nonlinear Langevin equation for a local growth of the profile is,

$$\frac{\partial h}{\partial t} = m\Delta^2 h + \frac{\lambda}{2}(\nabla h)^2 + \eta(x, t) \qquad \ldots (22)$$

The first term on the right-hand side describes relaxation of the interface by a tension term m. The second term is the lowest -order nonlinear term that can appear in the interface growth equation. Higher-order terms may also be present, but they are irrelevant and will not modify the scaling properties concerned. The noise $\eta(x, t)$

has a Gaussian distribution with $\langle \eta(x,t) \rangle = 0$ and $\langle \eta(x,t)\eta(x',t') \rangle = 2D\delta^d(x-x')\delta(t-t')$.

There is also a velocity term, but it is removed by choice of an appropriate moving coordinate system. Equation (22) is invariant under translation $h \rightarrow h + \text{constant}$, and obeys the infinitesimal reparametrization,

$h \rightarrow h + b. x, x \rightarrow x + \lambda bt,$

which describes the tilting of the interface by a small angle. When a given constituent point is added, the increment projected along the h-axis is,

$$\delta h = m\sqrt{[1 + (\nabla h)^2]} \simeq m + (m/2)(\nabla h)^2 + \cdots$$

Following the transformation $W(x,t) = e^{[(\lambda/2m)h(x,t)]}$, we have,

$$\frac{\partial W}{\partial t} = m\nabla^2 W + \left(\frac{\lambda}{2m}\right)\eta(x,t)W \qquad \qquad \dots(23)$$

which is a diffusion equation in a time-dependent random potential. $W(x,t)$ is the sum of Boltzmann weights for all static configurations of a flow in a $(d+1)$-dimensional space from $(0,0)$ to (x,t). The noise term describes a quenched random potential $(\lambda/2m)\eta(x,t)$ exerted by the environment. The second transformation, $m = -\nabla h$ results in

$$\frac{\partial m}{\partial t} + \lambda m . \nabla m = m\nabla^2 m - \nabla \eta(x,t) \qquad \qquad \dots(24),$$

which is the Burger's equation for a vorticity-free velocity field for $\lambda = 1$. In the Burger's equation, further evolution of the pattern proceeds through the larger parabolas growing at the expense of the smaller ones, and parallels the evolution of shock waves.

If the initial profile is $h(x,0) = h_0(x)$, its evolution is given by,

$h(x,t)$

$$= \frac{2m}{\lambda} \ln\left\{ \int_{-\infty}^{\infty} \frac{d^d\xi}{(4\pi mt)^{\frac{d}{2}}} e^{\left[-\frac{(x-\xi)^2}{2mt} + \frac{\lambda}{2m}h_0(\xi)\right]} \right\} \qquad \dots(25)$$

Let a given representation have bonds within itself, occupied by a resistance generated inside it due to its packing density Ø, with probability p. Let it have a support towards the concerned flow with probability $1 - p$. In such a representation, we have,

$$\left\langle \sum_k \varepsilon_k^n \right\rangle_{\xi,L} \propto L^{-x_n}, L \to \infty \qquad \qquad ...(26);$$

where $\langle \sum_k \varepsilon_k^n \rangle$ refers to the average over the sample realizations, L is the system-size $L \lesssim \xi$ and $\xi \propto (p - p_c)$ is the correlation length. ε_k is the energy dissipated in the branch k.

For a finite size scaling behaviour,

$$p\left(\sum_k \varepsilon_k^0, \sum_k \varepsilon_k^1, ..., \xi, L \right)$$

$$= \lambda^{x_0} \lambda^{x_1} ... p\left(\sum_k \varepsilon_k^0 / \lambda^{-x_0}, \frac{\sum_k \varepsilon_k^1}{\lambda^{-x_1}}, ..., \xi/\lambda, L/\lambda \right),$$

(λ is the rescaling parameter) equation (26) implies,

$$\left\langle \left(\sum_k \varepsilon_k^n \right)^m \right\rangle_{\xi,L} \propto L^{-mx_n} \qquad \qquad ...(27)$$

In disordered representations, the fluctuations of the free energy among different replicas may be regarded as the analogue of the temporal intermittency in a chaotic signal. Considering a spin-model of the D-dimensions, the Hamiltonian,

$$H[\{J_{ij}\}] = -\sum_{(i,j)} J_{ij} \sigma_i \sigma_j,$$

where $\sigma_i = \pm 1$ is the of the spin on the site i and the coupling J_{ij} is an independent random variable distributed according to a probability distribution $p(J_{ij})$. Given a coupling realization $\{J_{ij}\}$, the partition function of an N spin system is the trace of the Boltzmann weight $e^{(-\beta H_N)}$,

$$Z_N(\beta, \{J_{ij}\}) = \sum_{\{\sigma_i\}} e^{(-\beta H_N[\{J_{ij}\}])}$$

71

The free energy per spin in the limit $N \to \infty$ is,

$$F(\beta) = \lim_{N \to \infty} -\frac{1}{N\beta} \langle \ln Z_m \rangle$$

$$= \lim_{N \to \infty} -\frac{1}{N\beta} \int p(J_{ij}) dJ_{ij} \ln Z_N(\beta, \{J_{ij}\}) \qquad \ldots (28)$$

The free energy per spin of a coupling realization $\{J_{ij}\}$ of a N spin system is,

$$\Xi_N = -\frac{1}{N\beta} \ln Z_N(\beta, \{J_{ij}\})$$

The self-averaging of Ξ_N is

$$F = \lim_{N \to \infty} \Xi_N$$

For a unidimensional system with first neighbor interactions and uniform field h, we can write the partition function as the trace over 2×2 random transfer matrices product. The Hamiltonian is now $H = -\sum_i (J_i \sigma_i \sigma_{i+1} + h\sigma_i)$, such that,

$$Z_N = Tr \prod_{i=1}^{N} M_{i}, M_i$$

$$= \begin{pmatrix} e^{\beta J_i + \beta h} & e^{-\beta J_i + \beta h} \\ e^{-\beta J_i - \beta h} & e^{\beta J_i - \beta h} \end{pmatrix} \qquad \ldots (29)$$

The moments of the partition function can be estimated as an integral over the spectrum of the possible free energies $[\Xi_{min}, \Xi_{max}]$,

$$\langle Z_N(\beta)^q \rangle \propto \int \prod (\Xi) d\Xi \ e^{(-\beta \Xi q N)} \qquad \ldots (30)$$

The Kolmogorov entropy is related to the sum of the positive Lyapunov exponents which measure the divergence rate along the expanding directions, in accordance with the Theory of Physical Abstraction. For an ergodic measure with a compact support (as proved by Pesin) is,

$$K_1 \leq \sum_{i=1}^{P} \lambda_i;$$

where P is the number of exponents, $\lambda_i > 0$. In Hamiltonian systems,

$$K_1 = \sum_{i=1}^{p} \lambda_1 = \frac{dL^{(p)}}{dq}\Big|_{q=0}$$

A record of measures of a signal $x(t)$ at uniform spacing τ is

$$x_i = x(i\tau); i = 1,2, \dots, M \gg 1 \qquad \qquad \dots (31)$$

Clustering:

Since the number of eddies at scale l with singularity h is proportional to $l^{-d(h)}$, the number of grid points that have to be considered for resolving the set $S(h)$ is

$$N_h(R_e) \sim (L/\eta(h))^{d(h)} \propto R_e^{d(h)/(1+h)};$$

where $R_e = \dfrac{(\varepsilon L^4)^{1/3}}{\nu}$ and η is the dissipative Kolmogorov length.

Integrating over h, the total number of degrees of freedom is,

$$N(R_e) = \int dp(h)N_h(R_e) \propto R_e^{\delta};$$

where $\delta = \max_h [d(h)/1 + h]$.

The estimate $l_{min} = \eta(h_{min})$ assures that all the sets $S(h)$ are taken into account. The number of equations which allows us to get such a fully accurate description is thus;

$$N_T^* \sim \left(\frac{L}{l_{min}}\right)^3 \propto R_e^{3/(1+h_{min})};$$

which may be obtained by considering flows in the required number of directions or dimensions.

Let us define an effective mass dimension $\tilde{D}$ of the point on which the energy dissipation is concentrated by,

$$\tilde{h} = \frac{(\tilde{D} - 2)}{3}.$$

$D_1 \simeq 2.87$, which corresponds to select a $\tilde{h} = d\zeta_p/dP|_{p=3} \simeq 0.29$. $\tilde{l}$ is the smallest scale on which average active eddies are still present.

The minimum separation Δp between disturbances, with difference $\Delta\lambda$, is such that

$$\Delta p = Q\Delta\lambda \qquad \qquad ...(32);$$

$Q = 0.00642986 \times 10^5$ from experimental results.

Thus any number of points n inside a given representation will form a cluster point if they sufficiently close, as described by equation (32). The total information inside such a cluster is,

$$I_s = \frac{1}{n}\sum_{i=1}^{n} I_{s_i} \qquad \qquad ...(33);$$

where I_{s_i} is the energy dissipation information inside the constituent points. As the constituent points must be sufficiently close in order to form a cluster point, they may be considered to be a continuous energy dissipation information function, such that,

$$I_s = \frac{1}{n}\int_{1}^{n} I_s \, d\varepsilon$$

As I_s is a function of ε, we can write,

$$\int_{1}^{n} I_s \, d\varepsilon = \lim_{h\to 0} h \sum_{r=1}^{n} f(1+rh) \qquad \qquad ...(34);$$

where $nh = n - 1$, such that, $h = \frac{n-1}{n}$.

As $n \to \infty, h \to 0$.

A corollary to the Laws of Physical Transactions suggest that the individual constituent points of the cluster point will tend to be in the lowest effective dissipation energy state $\tilde{\varepsilon}$. Thus, the total energy dissipation information that the cluster point will tend to reach is,

$$\tilde{I_s} = nI_{\tilde{s}}$$

The total loss in energy dissipation energy information by the cluster point is therefore,

$$\Delta I_{\varepsilon} = I_{\varepsilon} - \tilde{I}_{\varepsilon} = \frac{1}{n} \sum_{i=1}^{n} I_{\varepsilon_i} - I_{\frac{1}{2}}$$

Let the individual energy dissipation information states of the constituent points be,

$$I_{\varepsilon_i} = \alpha_i I_{\frac{1}{2}},$$

such that the loss in dissipation energy information is,

$$\Delta I_{\varepsilon} = I_{\varepsilon} - \tilde{I}_{\varepsilon} = I_{\frac{1}{2}} \left[\left(\frac{1}{n} \sum_{i=1}^{n} \alpha_i \right) - 1 \right] \qquad \ldots (35)$$

Taking a continuous distribution,

$$\Delta I_{\varepsilon} = I_{\frac{1}{2}} \left[\frac{1}{n} \left(\int_{1}^{n} \alpha \, d\alpha \right) - 1 \right]$$

i.e.,

$$\Delta I_{\varepsilon} = I_{\frac{1}{2}} \left[n \left(\frac{n^2 - 1}{2} \right) - 1 \right] \qquad \ldots (36).$$

A close enough representation of the constituents of the universe therefore will be affected by this loss in dissipation energy information ΔI_{ε} inside its various clusters. This in turn will give rise to seemingly anomalous behaviour inside the representation. The clusters will seem to move away from each other with greater velocities than anticipated values. On the other hand, the clusters themselves will seem to be bound with greater strengths than is anticipated. The existence of dark energy and dark matter that we feel may be linked to the loss ΔI_{ε}.

Conclusion:

Choosing a suitable scaling ratio we may represent the microscopic and the macroscopic worlds on the same scale, enabling us to study and compare their various hitherto hidden properties. A large enough packing density ensures formation of cluster points inside the representations. These cluster points will tend to be in their lowest energy dissipation states. The whole universe being considered as a cluster when its constituents are close enough, as in the moment of the Big Bang, it tends to be in its lowest energy state (theoretically a zero energy state). ΔI_g, i.e., the difference in dissipation energy information which tends to infinity as the number of constituent points inside it tends to infinity, however establishes itself as Big Bang takes place. Yet, as the universe can expand further as its constituents move away from each other, with respect to a further expanded state it at any given present representation its clustered. Thus a hidden amount of energy dissipation information is present at any moment we look at the universe. This hidden energy dissipation information will make the clusters to move away from each other and the clusters themselves to be bound within themselves with greater hidden strengths than is anticipated.

8 More Discussions On The Topics Covered

Zero-postulation, as we can see, sits at the heart of abstraction. Null-postulation or zero-postulation favours no given result or a given set of results over all others. In that way, null postulation does not assume anything beforehand. What it does is to consider all possible results and derive the ultimate results from this exhaustive set of possible results. Each valid element inside the exhaustive set of results might interact in order to culminate into the 'real' results or happenings.

In not favouring solutions or sets of solutions, the principle of zero-postulation drives away any unwanted incompleteness from the description of the world. It is the interactions between the possible exhaustive set of solutions that creates the impression pointedness or directiveness in the universe, leading to the formation of clusters, as discussed earlier. These interactions may be chaotic in nature, giving rise to attractor points where the directiveness inside any given system asymptotically seem to approach. It is this directiveness, in turn, inside a given system or in the universe as a whole, that is the cause of all known phenomena. This directiveness of possibilities saves the universe from being exactly the same throughout, but makes it heterogeneously active, as we see it to be.

As zero-postulation considers the exhaustive set of all possible results, it would yield to be perfectly flexible to work with. That is to say, the scaling-ratio of observations may be adjusted as per requirements an intentions of the observer. This in turn is seen to unite observables at both microscopic and macroscopic levels through a similar basis of understanding. Even the 'non-real' abstract points themselves in the exhaustive set may be seen to interact and give rise to 'real' possibilities. Through analysis of all such possibilities, combined with analysis of all real elements inside the set is the description of the world in totality. The principle of null-postulation holds the key to such total description of the world, or of any given system, for that matter. Choosing suitable scaling-ratio and identifying all real and abstract parameters within a given system enables us to describe the system in all totality. Different scales of observations may have different sets of parameters with still different sets of interactions between themselves. These various levels of interactions between the parameters may give rise to different force-fields with their own respective sets of accelerations of interactions. However, all such fields of interactions, being fundamentally similar, has a similar basis of description.

In the description of the universe or any given system, there is always some 'hidden' information, that do not show up while the rest of the available information is being taken into account. Different sets of such available information at various levels of scaling will have their respective sets of such hidden information. This may prevent a given description of the world, or a part of it, at a given scale, from being completely deterministic. Lack of complete determinism in directiveness inside a given system, or in the universe as a whole, gives rise to hidden mass and hidden energy, that may not seem to show up in a given set of observations, but is seen to affect the overall description, nonetheless. This 'indeterminism' in directiveness will make the clusters to move away from each other and the clusters themselves to be bound within themselves with greater hidden strengths than is anticipated. The hidden mass exerts a hidden gravity inside the cluster concerned. The hidden energy, on the other hand, will tend to draw the clusters away from each other.

For a large enough or a small enough scale of observations, this hidden amount of information concerned seems to be of paramount significance. The galaxies themselves are seen to be held together by some hidden masses, while they are found to move away from each other by some hidden amount of energy. On the other hand, in the subatomic world, particles are seen to be held together by very strong hidden forces, while they seem to be interacting with each other with some hidden amount of energies.

In both the cases, a certain amount of hidden uncertainty in available information is always found to be present. At a scale of observations, which is more 'akin' to our own 'normal' day-to-day scale of looking at the world, which lies in between such vastness or such minuteness, the hidden amount of information seems to be of less significance. As the scaling-ratio and the object it is supposed to measure become increasingly commensurate to each other, the amount of hidden information for that given scale of measurements diminishes in size.

Looking at a large enough part of the universe, we may draw an analogy to a system of scattered particles, in motion or rest, relative to each other. These particles may or may not be similar to each other, if we look at a given locality. Our idea, however, is that we can always represent even the whole of the universe on a piece of paper of our desired size. We can very well do the same with localities of sub-atomic sizes.

We may represent both the worlds, viz., the microscopic and the macroscopic, within any desired standard size. Theoretically, we are only to diminish the snaps of the universe and magnify the snaps of the microscopic world in order to put both into representations of a definite scaling-size. Looking at such a representation of the macroscopic world (due to the large number of constituents and the large distances separating them involved) we will find it to be a complex mixture of various kinds of particles. On the other hand, looking at such a representation of the microscopic world, (due to the small distances separating the constituents) it will be like the actual universe itself, with various types of constituent parts involved. Such a representation of the microscopic and the macroscopic worlds will bring out hidden properties and behaviours of both worlds, as well as providing for a similar basis of studying them both.

From the Theory of Abstraction, we arrive at 'hidden' direction part of an energy-quantum. Quantum dynamics is seen to merge with classical dynamics if this hidden direction-part of the quantum-states are taken into consideration, as validated by practical analysis and data. Moreover, this hidden part of an actual energy-quantum may explain the dark-energy problem. As a support towards transport comprises the direction-part only, and as the resistance against motion is offered against the whole of an energy-quantum (direction-part + magnitude-part), this hidden energy may very well affect a gravitational field.

The relative periodicity of trajectories describing a given transaction seems to visit a given centre of possibilities or a given set of centers of possibilities. In the chaotic region, at least, any given event seems to have a periodic set of possibilities of happening. This implies the revisit of trajectories around one or a given set of attractor points. However, trajectories seem to stray from such points and not be approaching them exactly, for any given set of measurements. Adjusting respective scaling-ratios and choosing suitable parameters to a desired level of accuracy of predictions may ensure effective descriptions of a given system, a given set of systems, or the world as a whole.

Any change in scaling-ratio such that the constituents inside the system concerned have a large enough 'packing density', will yield the system, as a whole, to be in a 'condensed' state of being, at that

80

given scaling-ratio. Again, a suitable change in the scaling-ratio so as to break away from the condensation state yields a different set of observed results, complete with its own subset of hidden results and hidden information regions. This may explain how a whole world of its own set of possibilities can just come into being from such a condensed state of existence as the scaling-ratio is suitably adjusted. At such a scaling-ratio, this world is expected to have its own set of heterogeneity, which seems to be just the opposite of the practically homogenous level of existence as in the condensed state. Coming into existence of such a world, as well as, any non-existence of such a world may be attributed to changing scaling-ratios. Only a suitable change in the landscaping and scaling-ratio is sufficient in describing any such existence or non-existence, as a whole. This, in turn, may be taken as the explanation of how the whole universe can begin from a homogenous 'nothing' at the time of the 'Big Bang'.

www.ingramcontent.com/pod-product-compliance
Lightning Source LLC
Chambersburg PA
CBHW051220170526
45166CB00005B/1982